Undergraduate Lecture Notes in Physics (ULNP) publishes authoritative texts covering topics throughout pure and applied physics. Each title in the series is suitable as a basis for undergraduate instruction, typically containing practice problems, worked examples, chapter summaries, and suggestions for further reading.

ULNP titles must provide at least one of the following:

- An exceptionally clear and concise treatment of a standard undergraduate subject.
- A solid undergraduate-level introduction to a graduate, advanced, or non-standard subject.
- A novel perspective or an unusual approach to teaching a subject.

ULNP especially encourages new, original, and idiosyncratic approaches to physics teaching at the undergraduate level.

The purpose of ULNP is to provide intriguing, absorbing books that will continue to be the reader's preferred reference throughout their academic career.

More information about this series at http://www.springer.com/series/8917

Vittorio Degiorgio · Ilaria Cristiani

Photonics

A Short Course

Second Edition

Vittorio Degiorgio
Department of Electrical, Computer
and Biomedical Engineering
University of Pavia
Pavia
Italy

Ilaria Cristiani
Department Electrical, Computer
and Biomedical Engineering
University of Pavia
Pavia
Italy

ISSN 2192-4791 ISSN 2192-4805 (electronic)
Undergraduate Lecture Notes in Physics
ISBN 978-3-319-20626-4 ISBN 978-3-319-20627-1 (eBook)
DOI 10.1007/978-3-319-20627-1

Library of Congress Control Number: 2015947401

Springer Cham Heidelberg New York Dordrecht London

Printed on acid-free paper

Springer International Publishing AG Switzerland is part of Springer Science+Business Media
(www.springer.com)

Preface

Apart from a number of small corrections and updates, the main change with respect to the first edition is that a set of problems is provided at the end of each chapter, except the last one. The problems deal with numerical computations designed to illustrate the magnitudes of important quantities, and are also meant to reinforce the understanding of the material presented and to test the students' ability to apply theoretical formulas.

Pavia, Italy
May 2015

Vittorio Degiorgio
Ilaria Cristiani

Preface to the First Edition

The invention of the laser in 1960 has brought new ideas and new methods to many areas of science and technology. Starting from the consideration that, in a simplified view, a light beam can be seen as a stream of energy quanta, known as photons, the name "photonics" has been coined, in analogy to electronics, to indicate the generation and use of photon streams for engineering applications. The applications of photonics are everywhere around us from long-distance communications to DVD players, from industrial manufacturing to health-care, from lighting to image formation and display.

Many university programs in electrical engineering and applied physics propose courses on photonics. Our aim is to offer a concise, rigorous, updated book, that can serve as a textbook for an advanced undergraduate course. We believe that the book can be also useful for graduate students, professionals, and researchers who want to understand the principles and the functions of photonic devices.

The description of photonic devices essentially consists in a treatment of the interactions between optical waves and materials. When the interaction is non-resonant, an approach using Maxwell's equations and the macroscopic optical properties of materials is adequate. The description of optical amplification processes is more complex, because it becomes unavoidable to introduce quantum concepts. Our aim is to make laser action in atomic and semiconductor systems understandable also to an audience that has little background in quantum mechanics. Since photonics is based on a very fruitful interplay between physics and engineering, we have tried to keep the balance between these two aspects in all chapters. In order to give a feeling for the correct order of magnitudes of the discussed phenomena, several numerical examples are given throughout the book.

The book is divided into eight chapters. The first two chapters provide the background in electromagnetism and optics for the entire book. In particular, the second chapter describes the main optical components and methods of linear radiation–matter interactions. The third chapter covers the principles of lasers, and the properties of continuous and pulsed laser sources. Chapters 4 and 7 deal with nonlinear interactions that produce processes such as modulation, switching, and frequency conversion of optical beams. Chapter 5 describes photonic devices based on semiconductors, including not only lasers, but also light-emitting diodes and photodetectors. The properties of optical fibers, and of different fiber-optic components and devices are the subject of Chap. 6. The last chapter discusses the most important applications, emphasizing the main principles, while avoiding detailed technical descriptions that may quickly become obsolete in this rapidly changing field. Introductory information is also given on arguments of particular practical importance, such as solid-state lighting, displays, and photovoltaic cells.

It should be emphasized that the invention of the laser has deeply affected many fields of research, either by putting topics, such as optical spectroscopy, light scattering, and coherence theory on a completely different footing, or by creating entirely new topics, such as the physics of coherent light sources, nonlinear optics and quantum optics. Our text does not touch upon those new topics that generally require a deeper understanding of quantum concepts. The only exception is nonlinear optics, which can be mostly discussed in classical terms.

The book systematically uses the SI metric system with the exception of the use of the electron volt (eV) for the energy of photons and atomic levels. The system of units is summarized in Appendix A, together with a list of values of some fundamental physical constants. Use of acronyms is kept to a minimum, and the list of those used is given in Appendix B.

The book is intended to be self-contained. The prerequisites include a background knowledge from standard courses in general physics, electromagnetic fields, and basic electronics. Some background in quantum physics would be helpful. Considering the level of treatment, we have judged it not useful to provide references to articles that have appeared in scientific journals. Instead, to help an interested reader gain a deeper insight into the fundamentals of photonics or into a particular application, a bibliography containing a selection of a few excellent reference books is given.

We are especially indebted to Lee Carroll for providing us with many suggestions that greatly improved the presentation. We also acknowledge useful conversations with Antoniangelo Agnesi, Cosimo Lacava, Roberto Piazza, and Giancarlo Reali.

Pavia, Italy
February 2014

Vittorio Degiorgio
Ilaria Cristiani

Contents

Chapter 1
Electromagnetic Optics

Abstract Light is an electromagnetic phenomenon that is described, in common with radio waves, microwaves and X-rays, by Maxwell equations. This chapter briefly reviews those aspects of electromagnetic theory that are particularly relevant for optics. The propagation of quasi-collimated monochromatic light beams, Gaussian waves, is treated by using the paraxial approximation. Diffraction theory is described as an approximate theory of wave propagation. In particular the Fresnel approach is shown to be equivalent to the paraxial picture. Finally, the use of the Fraunhofer approximation allows establishing a very useful relation between the far-field pattern of the diffracted field and the spatial Fourier transform of the field distribution immediately after the screen plane.

1.1 Spectrum of Electromagnetic Waves

The typical nomenclature of electromagnetic waves is given in Table 1.1, along with the wavelength ranges associated with the different spectral bands. The field of Photonics covers the infrared, visible, and ultraviolet wavelength ranges.

The term "light" is born to indicate electromagnetic radiation having wavelength within the interval 0.4–0.75 μm, which is the interval spanned by the radiation emitted by the sun. As a result of biological adaptation, this is also the wavelength range to which the retina of our eyes is sensitive. In contemporary scientific terminology, the term "light" is also applied to infrared and ultraviolet radiation.

In this text ν expresses the frequency of light, and is measured in units of hertz (Hz). Alternatively, $\omega = 2\pi\nu$, the angular frequency of light can be used, and is measured in radians per second. The wavelength of light, expressed as λ, is usually measured in micrometers/microns ($1\ \mu\text{m} = 10^{-6}\,\text{m}$) or nanometers ($1\ \text{nm} = 10^{-9}\,\text{m}$). The speed of light in vacuum is expressed as c, and is related to the wavelength and frequency by $c = \lambda\nu$. In general, the International System (SI) of units will be used. Appendix A gives the list of prefixes used to indicate multiples and submultiples of the fundamental units.

Table 1.2 shows the wavelength ranges corresponding to different colors of visible electromagnetic waves.

V. Degiorgio and I. Cristiani, *Photonics*, Undergraduate Lecture Notes in Physics,
DOI 10.1007/978-3-319-20627-1_1

Table 1.1 Spectrum of electromagnetic waves

Name	Frequency	Wavelength
Ultralow frequencies	<30 kHz	>10 km
Low frequencies	30–300 kHz	10–1 km
Radio frequencies	0.3–3 MHz	1–0.1 km
High frequencies (HF)	3–30 MHz	100–10 m
VHF	30–300 MHz	10–1 m
UHF	0.1–1 GHz	1–0 .3 m
Microwaves	1–100 GHz	30–0.3 cm
Millimeter waves	0.1–3 THz	3–0.1 mm
Infrared	3–400 THz	100–0.75 μm
Visible	400–750 THz	0.75–0.4 μm
Ultraviolet	750–1,500 THz	0.4–0.2 μm
Vacuum UV	1.5–6 $\times 10^3$ THz	200–50 nm
Soft X rays	6–300 $\times 10^3$ THz	50–1 nm
X rays	0.3–30 $\times 10^6$ THz	1–0.01 nm
γ rays	$>10^6$ THz	<0.3 nm

Table 1.2 Visible electromagnetic waves

Color	λ (nm)
Violet	400–450
Blue	450–495
Green	495–575
Yellow	575–595
Orange	595–620
Red	620–750

Historically, the nature of light has been subject to considerable debate. Some scientists, like Huygens and Young, posited a "wave-like" nature of light, while others, like Newton, proposed a "corpuscular" point of view. The highly successful description of light provided by Maxwell's equations seemed to end the debate, by providing a strong validation of the wave-like interpretation. However the controversy reopened in the last part of the nineteenth century, because the wave-like interpretation was inconsistent with experimental observations of certain phenomena such as the photoelectric effect and the spectral radiance of blackbody emission. Planck and Einstein clearly demonstrated that electromagnetic energy is emitted and absorbed in discrete units, or fundamental quanta of energy, now called photons.

Quantum theory provides a unified view that combines both the wave-like and the corpuscular aspects of light. Traditionally, textbooks discussing lasers and related applications do not make full use of the quantum theory of light. Instead, they adopt an intermediate approach by treating all wave propagation phenomena with Maxwell's equations, and introducing light quanta only when discussing absorption and emission phenomena. This intermediate approach is also used in this book, as it is particularly suited to a concise introduction to Photonics.

Considering a monochromatic plane wave with frequency ν and wave vector $\mathbf{k}$, the corresponding photon is a quasi-particle without mass, with an energy of $h\nu$ and a momentum of $h\mathbf{k}/(2\pi)$. The universal constant, h, called Planck constant, has value 6.55×10^{34} J·s. Usually the photon energy is conveniently expressed in units of electronvolt (eV): 1 eV is the energy acquired by an electron (electric charge $e = 1.6 \times 10^{-19}$ C) when it is accelerated by a potential difference of 1 V, and so $1\,\text{eV} = 1.6 \times 10^{-19}$ J.

1.2 Electromagnetic Waves in Vacuum

The vacuum propagation of electromagnetic waves is described by Maxwell's equations:

$$\nabla \cdot \mathbf{E} = 0, \tag{1.1}$$

$$\nabla \cdot \mathbf{B} = 0, \tag{1.2}$$

$$\nabla \times \mathbf{B} = \varepsilon_o \mu_o \frac{\partial \mathbf{E}}{\partial t}, \tag{1.3}$$

$$\nabla \times \mathbf{E} = -\frac{\partial \mathbf{B}}{\partial t}, \tag{1.4}$$

where $\mathbf{E}$ is the electric field (measured in V/m), and $\mathbf{B}$ is the magnetic induction (measured in T). The constant $\varepsilon_o = 8.854 \times 10^{-12}$ F/m is called the dielectric constant (or electric permittivity) of free space, and $\mu_o = 4\pi \times 10^{-7}$ H/m is the vacuum magnetic permeability.

Recalling a general property of differential operators:

$$\nabla \times (\nabla \times \mathbf{E}) = \nabla(\nabla \cdot \mathbf{E}) - \nabla^2 \mathbf{E}, \tag{1.5}$$

taking the rotor of both members of (1.4), and using (1.1), it is found that:

$$\nabla \times (\nabla \times \mathbf{E}) = -\nabla^2 \mathbf{E} = -\frac{\partial(\nabla \times \mathbf{B})}{\partial t}. \tag{1.6}$$

By using (1.3), one obtains a single equation, called the wave equation, in which the unknown function is the electric field:

$$\nabla^2 \mathbf{E} = \frac{1}{c^2} \frac{\partial^2 \mathbf{E}}{\partial t^2}, \tag{1.7}$$

where $c = (\varepsilon_o \mu_o)^{-1/2}$ is the propagation velocity of the electromagnetic wave. It can be shown that $\mathbf{B}$ satisfies an equation that is identical to (1.7).

The wave described by (1.7) transports both energy and momentum. The flow of electromagnetic energy is governed by the Poynting vector, defined as:

$$\mathbf{S_P} = \mathbf{E} \times \mathbf{H}, \tag{1.8}$$

where $\mathbf{H} = \mathbf{B}/\mu_o$ is the magnetic field (units: A/m). The modulus of $\mathbf{S_P}$, which has dimension W/m^2, represents the power per unit area transported by the electromagnetic wave.

An important particular solution of (1.7) is the monochromatic plane wave:

$$\mathbf{E} = \mathbf{E_o} \cos(\omega t - \mathbf{k} \cdot \mathbf{r} + \phi) = \mathbf{E_o} Re\{\exp[-i(\omega t - \mathbf{k} \cdot \mathbf{r} + \phi])\}, \tag{1.9}$$

where $\mathbf{k}$ is the propagation vector, with modulus $k = 2\pi/\lambda = \omega/c$, and ϕ is a constant phase that can always be put equal to 0 by an appropriate choice of the time origin. The wave described by (1.9) generates plane equal-phase surfaces (also called wavefronts) that are perpendicular to $\mathbf{k}$.

By substituting (1.9) into (1.1), one finds: $i\mathbf{k} \cdot \mathbf{E} = 0$, that is: $\mathbf{E} \perp \mathbf{k}$. Analogously, from (1.2) one derives $\mathbf{B} \perp \mathbf{k}$. In addition, Maxwell equations say that $\mathbf{E} \perp \mathbf{B}$, that $\mathbf{E}$ and $\mathbf{B}$ have the same phase, and that $B = \sqrt{\varepsilon_o \mu_o}\, E$. Summarizing, the electromagnetic plane wave is a transverse wave, in which electric and magnetic field vectors are mutually orthogonal and lie in a plane that is perpendicular to the propagation direction.

The Poynting vector of the plane wave is parallel to the propagation vector $\mathbf{k}$. By inserting (1.9) into (1.8), the following expression is found:

$$\mathbf{S_P} = c\varepsilon_o E_o^2 \frac{\mathbf{k}}{k} \cos^2(\omega t - \mathbf{k} \cdot \mathbf{r} + \phi). \tag{1.10}$$

Optical fields oscillate at very high frequencies. For instance, at the wavelength $\lambda = 1\,\mu$m, the corresponding oscillation period is $T = \lambda/c = 3.3 \times 10^{-15}$ s $= 3.3$ fs. In most cases it is more significant to consider time averaged quantities instead of instantaneous values. To this end, it is useful to recall a general property of sinusoidal functions: given two functions oscillating at the same frequency, $a(t) = A\cos(\omega t + \phi_A)$ and $b(t) = B\cos(\omega t + \phi_B)$, the time average of the product is:

$$\langle a(t)b(t) \rangle = \frac{1}{2} AB \cos(\phi_A - \phi_B). \tag{1.11}$$

Using (1.11), it is found that the time-average of the modulus of $\mathbf{S_P}$, which is the intensity of the wave, is given by:

$$I = c\frac{\varepsilon_o E_o^2}{2}. \tag{1.12}$$

As shown in (1.9), the electric field can be described by using complex exponentials of the type $\mathbf{E} = \mathbf{E_o}\exp[-i(\omega t - \mathbf{k}\cdot\mathbf{r} + \phi)]$ instead of real sinusoidal functions. In this text the exponential notation will be often used to simplify the calculations. In any case it should never be forgotten that only the real part of the complex exponential corresponds to the real electric field.

1.3 Polarization of Light

The direction of polarization of an electromagnetic wave is the direction of the vector $\mathbf{E}$. Assume that a plane wave is propagating in the z direction. The electric field is oriented in the x-y plane. Its x and y components are both sinusoidal functions of time:

$$E_x = E_{xo}\cos(\omega t - kz), \tag{1.13}$$

$$E_y = E_{yo}\cos(\omega t - kz + \phi), \tag{1.14}$$

where ϕ is the phase shift between the two components. Putting $X = E_x/E_{xo} = \cos(\omega t - kz)$ and $Y = E_y/E_{yo} = \cos(\omega t - kz + \phi) = \cos(\omega t - kz)\cos\phi - \sin(\omega t - kz)\sin\phi$, it is possible to write an implicit equation containing X and Y, but not the spatio-temporal coordinates:

$$X\cos\phi - Y = \sqrt{1 - X^2}\sin\phi. \tag{1.15}$$

By squaring both sides of (1.15) one obtains:

$$X^2 + Y^2 - 2XY\cos\phi = \sin^2\phi, \tag{1.16}$$

hence:

$$\frac{E_x^2}{E_{xo}^2} + \frac{E_y^2}{E_{yo}^2} - 2\frac{E_x E_y}{E_{xo}E_{yo}}\cos\phi = \sin^2\phi. \tag{1.17}$$

Equation (1.17) describes the trajectory of the endpoint of the vector $\mathbf{E}$ in the plane x-y. The motion is periodic in time, with period $T = 1/\nu$. In the general case, the trajectory is an ellipse, as shown in Fig. 1.1, and the wave is said to be elliptically polarized. In the particular case of $\phi = 0$, the ellipse reduces to a straight line, i.e., the direction of the electric field vector does not change with time. The straight line forms an angle α with the x axis, given by the relation $\tan(\alpha) = E_{yo}/E_{xo}$. Linear polarization is also obtained when $\phi = \pi$, in this case the polarization direction forms an angle $-\alpha$ with the y axis.

In the case $\phi = \pm\pi/2$ and $E_{xo} = E_{yo}$, the endpoint of the vector $\mathbf{E}$ sketches out a circle in one optical period, and so the wave is said to be circularly polarized. The rotation of $\mathbf{E}$ can be clockwise, if $\phi = \pi/2$, or anti-clockwise, if $\phi = -\pi/2$.

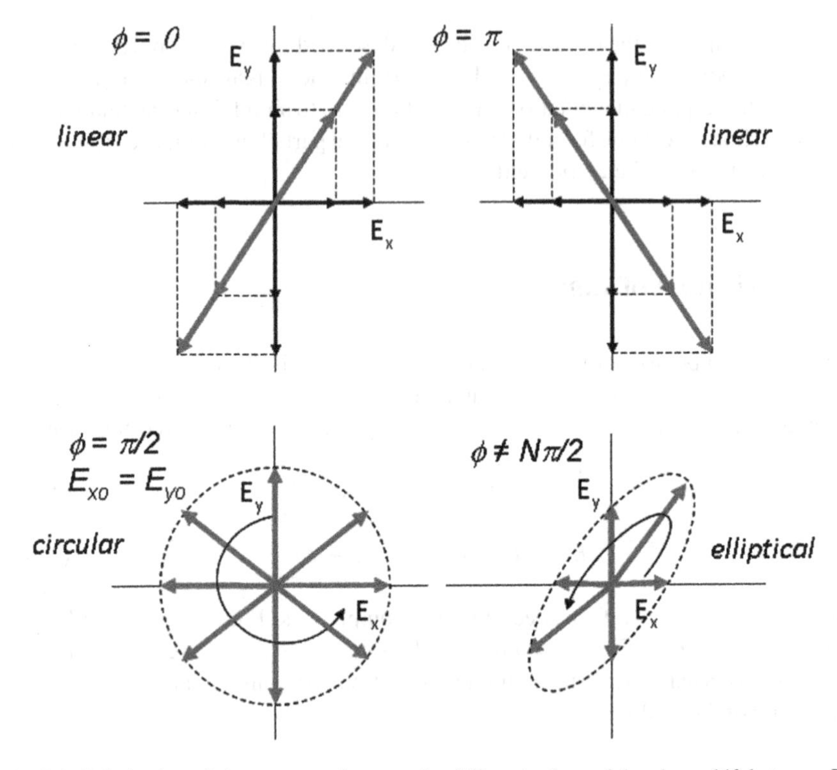

Fig. 1.1 Polarization of electromagnetic waves for different values of the phase shift between E_x and E_y

A generic polarization state can also be described as a superposition of two circular polarizations, each having a different amplitude and phase. If the two amplitudes are different the resulting polarization is elliptical. If the amplitudes are equal, the superposition corresponds to a linear polarization, having direction controlled by the phase shift between the two circular polarizations.

It is usually said that the light emitted by the Sun and by conventional sources is unpolarized. What is meant here is that the direction of the electric field observed in a given point is an essentially random function of time

1.4 Paraxial Approximation

Representing the field of the monochromatic wave as: $\mathbf{E}(\mathbf{r}, t) = \mathbf{E}(\mathbf{r})\exp(-i\omega t)$, and substituting this expression into the wave equation (1.7), allows the so-called Helmholtz equation to be obtained:

$$(\nabla^2 + k^2)\mathbf{E}(\mathbf{r}) = 0 \tag{1.18}$$

In real situations one never deals with ideal plane waves, but rather with collimated beams, like those emitted by lasers. It is therefore useful to develop approximated treatments applicable to light beams possessing a relatively small distribution of wave vectors. Since the discussion will only concern the spatial properties of monochromatic fields, the time-dependent term $\exp(-i\omega t)$ will be omitted from now on in this chapter inside all the expressions involving the electric field.

Considering a light beam that travels along the z axis, and substituting the following expression for the electric field inside (1.18):

$$E(x, y, z) = U(x, y, z)\exp(ikz), \tag{1.19}$$

the Helmholtz equation becomes:

$$\frac{\partial^2 U}{\partial x^2} + \frac{\partial^2 U}{\partial y^2} + \frac{\partial^2 U}{\partial z^2} + 2ik\frac{\partial U}{\partial z} = 0. \tag{1.20}$$

It is now assumed that U is a slowly varying function of z, where slowly varying means that the spatial scale over which U has significant variations is much bigger than the wavelength λ, as qualitatively shown in Fig. 1.2. Mathematically, this is equivalent to state that:

$$\frac{dU}{dz} \ll kU. \tag{1.21}$$

A consequence of (1.21) is that the second derivative of U with respect to z is much smaller than kdU/dz, and so it can be neglected inside (1.20), obtaining the so-called paraxial equation:

$$\frac{\partial^2 U}{\partial x^2} + \frac{\partial^2 U}{\partial y^2} + 2ik\frac{\partial U}{\partial z} = 0. \tag{1.22}$$

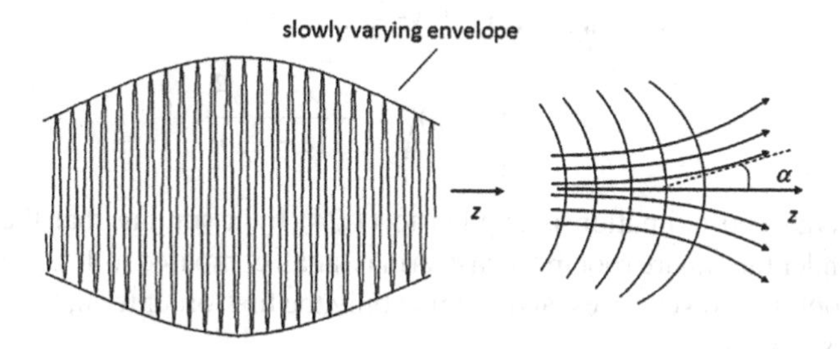

Fig. 1.2 Paraxial approximation

In order to better clarify the meaning of the paraxial approximation, it is useful to consider the simple case of a wave made by the superposition of three plane waves, one having vector **k** directed along the z axis, the two others with **k** laying in the plane x-z and forming angles $\pm\alpha$ with the z axis. By assuming that the three plane waves have the same amplitude U_o, $U(x, y, z)$ can be written as:

$$U(x, y, z) = U_o\{1 + \exp[ik(\cos\alpha - 1)z][\exp(ikx \sin\alpha) + \exp(-ikx \sin\alpha)]\}.$$

The first derivative with respect to z is:

$$\frac{dU}{dz} = ik(\cos\alpha - 1)(U - U_o).$$

If α is small, then the quantity $\cos\alpha - 1$ is also small with respect to 1. Therefore it is demonstrated that:

$$\left|\frac{dU}{dz}\right| \ll |k(U - U_o)| < |kU|.$$

1.4.1 Spherical Waves

An important particular solution of the wave equation (1.7) is a wave with spherical wavefronts, whose electric field, in scalar form, is expressed as:

$$E = \frac{A_o}{|\mathbf{r} - \mathbf{r}_o|}\exp(ik|\mathbf{r} - \mathbf{r}_o|) \tag{1.23}$$

The origin of the spherical wave is at point $\mathbf{r_o}$ having coordinates x_o, y_o, z_o. Putting the origin on the z axis means $x_o = y_o = 0$, and the distance $|\mathbf{r} - \mathbf{r_o}|$ is given by:

$$\begin{aligned} |\mathbf{r} - \mathbf{r_o}| &= \sqrt{x^2 + y^2 + (z - z_o)^2} \\ &= (z - z_o)\sqrt{1 + \frac{x^2 + y^2}{(z - z_o)^2}} \end{aligned} \tag{1.24}$$

The paraxial approximation is applied to (1.24) by assuming that the fraction appearing under the square root in the last term is small compared to 1. By expanding the square root in a power series, and keeping only the first-order term, the following expression is derived:

$$|\mathbf{r} - \mathbf{r_o}| \approx (z - z_o)\left[1 + \frac{x^2 + y^2}{2(z - z_o)^2}\right]. \tag{1.25}$$

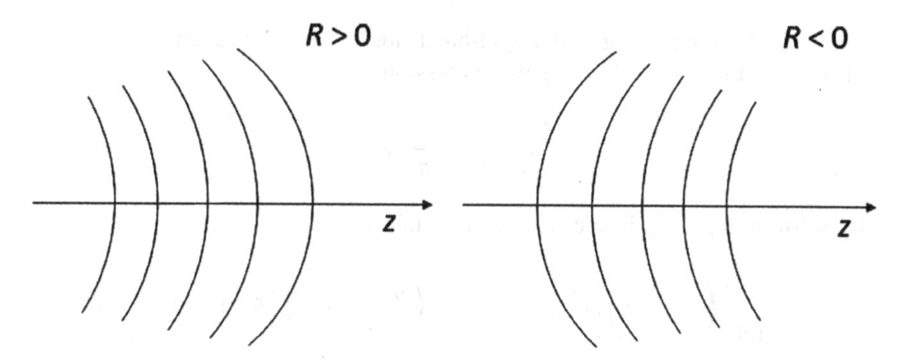

Fig. 1.3 Convention for the sign of the radius of curvature of the spherical wave

To use (1.25) instead of (1.24) is analogous to approximating the spherical wavefront with a paraboloid. The validity of (1.25) is clearly limited to situations in which only that part of the wavefront for which the distance from the z axis is much less than $z - z_o$ is considered.

Calling $R(z) = z - z_o$ the radius of curvature of the wavefront allows the complex amplitude of the electric field associated to the spherical wave to be written as:

$$U(x, y, z) = \frac{A_o}{R(z)} \exp\left(ik\frac{x^2 + y^2}{2R(z)}\right). \tag{1.26}$$

The radius of curvature can be positive or negative: the chosen convention (shown in Fig. 1.3) being that positive R indicates a wavefront having convexity oriented toward positive z, i.e. when the wave is diverging.

1.4.2 Gaussian Spherical Waves

As will be shown in Chap. 3, the beam emitted by a laser can usually be described as a spherical wave presenting a Gaussian amplitude distribution in the x-y plane. The electric field of this Gaussian spherical wave is:

$$\begin{aligned} U(x, y, z) &= A(z) \exp\left[-\frac{x^2 + y^2}{w^2}\right] \exp\left[i\frac{k(x^2 + y^2)}{2R}\right] \\ &= A(z) \exp\left[i\frac{k(x^2 + y^2)}{2q}\right], \end{aligned} \tag{1.27}$$

where w is the distance from the z axis at which the field amplitude is reduced by a factor $1/e$. The parameter w can be called radius of the Gaussian wave.

The quantity q appearing in the right-hand side of (1.27) is called the complex radius of curvature. It is defined by the expression:

$$\frac{1}{q} = \frac{1}{R} + i\frac{\lambda}{\pi w^2} \tag{1.28}$$

By substituting (1.27) inside (1.22), one obtains:

$$\left[\frac{k^2}{q^2}\left(\frac{dq}{dz} - 1\right)(x^2 + y^2) + \frac{2ik}{q}\left(\frac{q}{A}\frac{dA}{dz} + 1\right)\right]A(z) = 0 \tag{1.29}$$

Equation (1.29) can be satisfied for all possible values of x and y provided that:

$$\frac{dq}{dz} = 1\ ; \qquad \frac{dA(z)}{dz} = -\frac{A(z)}{q(z)}. \tag{1.30}$$

By using the initial conditions $q(z_o) = q_o$, $A(z_o) = A_o$, the following solutions to (1.30) are found:

$$q(z) = q_o + z - z_o\ ; \qquad \frac{A(z)}{A_o} = \frac{q_o}{q(z)}. \tag{1.31}$$

It is therefore demonstrated that the Gaussian spherical wave is a solution of the paraxial equation. This means that the wave does not change its functional form under propagation. What happens is that the amplitude $A(z)$ and the complex radius of curvature $q(z)$ change with z following simple algebraic laws. In particular the evolution of $q(z)$ is identical to that of the radius of curvature of the spherical wave.

The case of a Gaussian spherical wave having a planar wavefront at $z = 0$ is now considered. If the radius of curvature of the wavefront is infinite for $z = 0$, then the reciprocal of the complex radius of curvature at $z = 0$, as derived from (1.28), is: $1/q_o = i\lambda/(\pi w_o^2)$, where w_o is the beam radius at $z = 0$. The field distribution at $z = 0$ is:

$$U(x_o, y_o, 0) = A_o \exp\left[i\frac{k(x_o^2 + y_o^2)}{2q_o}\right] = A_o \exp\left[-\frac{x_o^2 + y_o^2}{w_o^2}\right]. \tag{1.32}$$

Having assigned $U(x_o, y_o, 0)$, the field distribution at the generic coordinate z is immediately derived by using the two conditions of (1.31) that describe how the amplitude and the complex radius of curvature are modified during the propagation along the z axis. Note that the propagation problem is solved through the algebraic relations (1.31) without making use of the differential equation (1.22). It is found:

$$U(x, y, z) = A_o \frac{q_o}{q_o + z} \exp\left[i\frac{k(x^2 + y^2)}{2(q_o + z)}\right], \tag{1.33}$$

which can also be written as:

$$U(x, y, z) = A_o \frac{w_o}{w(z)} \exp[-i\psi(z)] \exp\left[i\frac{k(x^2+y^2)}{2q(z)}\right], \tag{1.34}$$

where

$$q(z) = z + q_o = z - \frac{i\pi w_o^2}{\lambda}, \tag{1.35}$$

$$w(z) = w_o\sqrt{1+\left(\frac{\lambda z}{\pi w_o^2}\right)^2} = w_o\sqrt{1+\left(\frac{z}{z_R}\right)^2}, \tag{1.36}$$

and

$$\psi(z) = \arctan\left(\frac{\lambda z}{\pi w_o^2}\right) = \arctan\left(\frac{z}{z_R}\right). \tag{1.37}$$

The quantity z_R is called Rayleigh length, and is defined as follows:

$$z_R = \frac{\pi w_o^2}{\lambda}. \tag{1.38}$$

The behavior of the beam radius $w(z)$ is reported in Fig. 1.4. It is seen that the Gaussian spherical wave has the minimum radius at the position (usually called beam waist) in which its wavefront is planar. The Rayleigh length represents the propagation distance at which the beam radius becomes larger than w_o by a factor $\sqrt{2}$. The beam radius is slowly growing for $z \leq z_R$ and tends to grow linearly as $z \gg z_R$.

Equation (1.34) indicates that the propagating Gaussian spherical wave acquires, with respect to the plane wave, an additional phase delay of $\psi(z)$ that, according to (1.37), varies from $-\pi/2$ for $z \ll z_R$ to $\pi/2$ for $z \gg z_R$, as shown in Fig. 1.5.

By using (1.28) and (1.35) the z-dependence of the radius of curvature of the wavefront is derived:

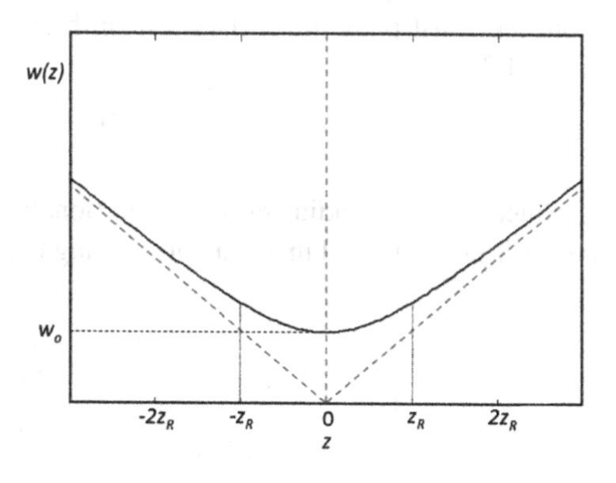

Fig. 1.4 Gaussian spherical wave: behavior of the beam radius, $w(z)$, as a function of the propagation distance

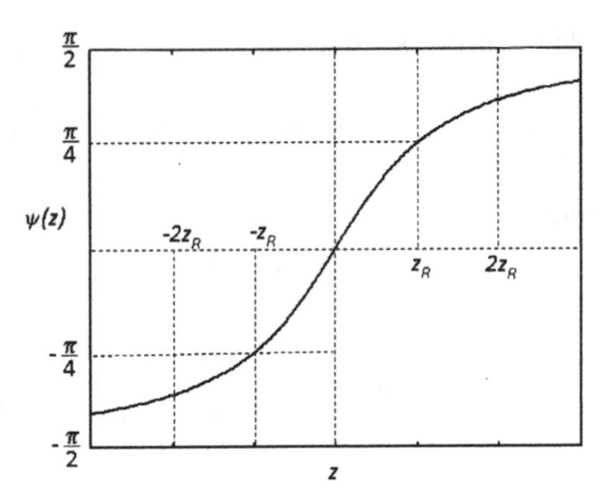

Fig. 1.5 Gaussian spherical wave: behavior of the additional phase delay, $\psi(z)$, as a function of the propagation distance

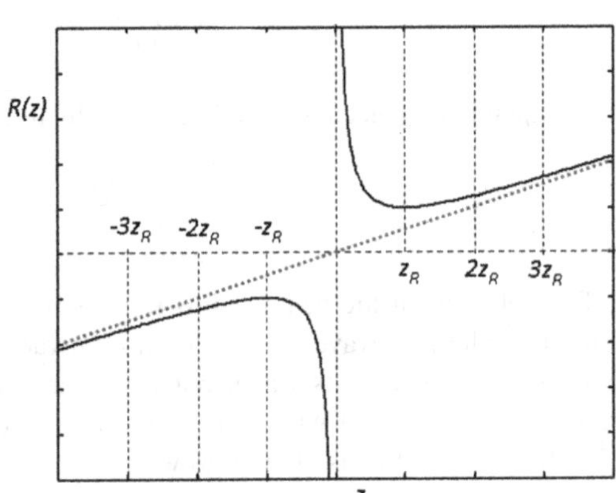

Fig. 1.6 Gaussian spherical wave: behavior of the radius of curvature of the wavefront, $R(z)$, as a function of the propagation distance

$$R(z) = z + \frac{1}{z}\left(\frac{\pi w_o^2}{\lambda}\right)^2 = z + \frac{{z_R}^2}{z}. \tag{1.39}$$

The behavior of $R(z)$ is illustrated in Fig. 1.6.

It is interesting to discuss the behavior of the Gaussian spherical wave at distances from the waist plane much larger than the Rayleigh length. If $z \gg z_R$, one finds from (1.36):

$$w(z) \approx \frac{\lambda z}{\pi w_o}. \tag{1.40}$$

Since the beam radius grows proportionally to distance, most of the propagating power is concentrated inside a cone having the aperture:

$$\theta_o = \frac{\lambda}{\pi w_o}. \tag{1.41}$$

The angle θ_o is the divergence angle of the beam. Equation (1.41) expresses a very general concept for wave propagation: the finite transversal size of the beam is necessarily associated to an angular spread of the wave vector directions. Apart from a numerical factor of the order of one, the divergence angle is given by the ratio between wavelength and transversal size of the light beam. As it will be seen in the following sections, the same result is found in the classical diffraction problem, in which a plane wave crossing a screen with a circular aperture of radius $D/2$ generates a beam with an angular spread $\approx 2\lambda/D$.

To complete the discussion concerning the behavior at $z \gg z_R$, (1.39) shows that $R(z)$ approaches the value of z, that is, it tends towards the value expected for a spherical wave centered at $z = 0$.

The paraxial equation admits other beam-like solutions that, like the spherical Gaussian wave, do not change their shape during free propagation. An interesting family of such solutions is that of Hermite-Gauss spherical waves, whose amplitude, in Cartesian coordinates, is the product of a Gaussian function times two Hermite polynomials:

$$U(x, y, z) = A_o \frac{w_o}{w(z)} \exp\left[-i(l + m + 1)\psi(z)\right]$$
$$\exp\left[i \frac{k(x^2 + y^2)}{2q(z)}\right] H_l\left(\sqrt{2}\frac{x}{w(z)}\right) H_m\left(\sqrt{2}\frac{y}{w(z)}\right), \tag{1.42}$$

where the Hermite polynomial of order n is defined as

$$H_n(\xi) = \exp(\xi^2)(-1)^n \frac{d^n}{d\xi^n} \exp(-\xi^2). \tag{1.43}$$

Applying (1.43) to generate, as an example, the polynomials of order 0, 1, 2, it is found:

$$H_0(\xi) = 1; \quad H_1(\xi) = 2\xi; \quad H_2(\xi) = 4\xi^2 - 2. \tag{1.44}$$

The Hermite-Gaussian beam of order (0,0) is the simple Gaussian beam. As will be seen in Chap. 3, the Hermite-Gaussian beams are useful to describe the field distribution inside laser cavities.

To summarize, the treatment developed in this section has demonstrated that Gaussian spherical waves have propagation properties similar to those of spherical waves in the geometrical optics approach. However one important difference is that all the formulas developed in this section include the effects due to diffraction.

1.5 Diffraction Fresnel Approximation

The classical diffraction theory concerns the situation illustrated in the scheme of Fig. 1.7. A monochromatic plane wave, with propagation vector **k** parallel to the z axis is transmitted through an aperture Σ in an opaque screen laying on plane x-y. In

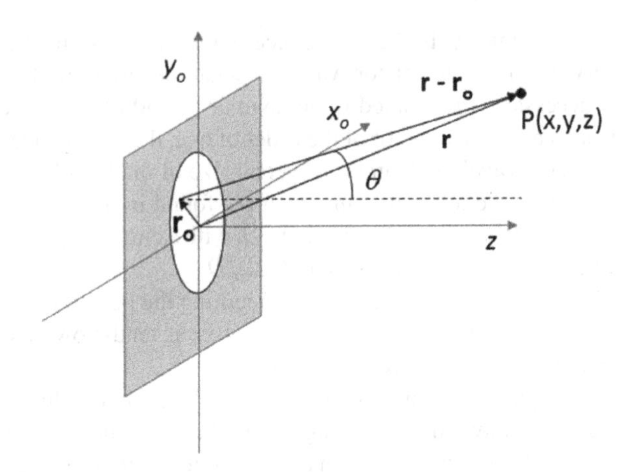

Fig. 1.7 Diffraction from an aperture Σ in the plane x-y

order to exactly determine the spatial distribution of the electromagnetic field in the semi-space beyond the screen, one should solve the wave equation with appropriate boundary conditions. In most cases the exact calculations are too complicated, and so approximate theories have to be developed. The simplest approach would be that of geometrical optics: the electric field of the transmitted wave would be different from zero only in the cylindrical region having axis parallel to z and the aperture Σ as the basis. Since it is experimentally observed that the pattern of the transmitted wave extends outside the simple shadow of the aperture, it is necessary to go beyond this geometrical optics approach. The standard diffraction theory, put forward by Huygens and successively elaborated by Fresnel and Kirchhoff, is an approximate theory based on the following model. Each infinitesimal element of Σ generates a spherical wave, having amplitude and phase determined by the incident field. The electric field at an observation plane at distance z from the screen is given by the superposition of all the spherical waves coming from the different infinitesimal elements composing the aperture Σ. The phase of each spherical wave is fixed by the distance between the considered surface element and the observation point. Note that the superposition principle is direct consequence of the linearity of Maxwell equations.

The incident field is not necessarily a plane wave. It can be generally described by a complex amplitude $E_i(x_o, y_o)$, where x_o, y_o are the coordinates on the screen plane. The diffracted field is given at point $P(x, y, z)$ by the following surface integral:

$$E(\mathbf{r}) = \frac{i}{\lambda} \int \int_{\Sigma} E_i(\mathbf{r}_o) \frac{\exp(ik|\mathbf{r} - \mathbf{r_o}|)}{|\mathbf{r} - \mathbf{r}_o|} \cos\theta dx_o dy_o, \tag{1.45}$$

where the vectors $\mathbf{r} = (x, y, z)$ and $\mathbf{r}_o$ represent, respectively, the position of P and the position of the surface element $dx_o dy_o$. θ is the angle formed by the vector $\mathbf{r} - \mathbf{r}_o$ with the z axis. Therefore:

$$\cos\theta = \frac{z}{\sqrt{(x - x_o)^2 + (y - y_o)^2 + z^2}}, \tag{1.46}$$

and

$$|\mathbf{r}-\mathbf{r}_o| = \sqrt{(x-x_o)^2+(y-y_o)^2+z^2} = z\sqrt{1+\frac{(x-x_o)^2+(y-y_o)^2}{z^2}}. \tag{1.47}$$

The imaginary unit i placed in front of the integral (1.45) indicates that the spherical wave coming from the generic surface element is phase-shifted by $\pi/2$ with respect to the incident wave. Provided that the diffracted field is evaluated at an observation point satisfying the condition $z \gg \lambda$, the integral (1.45) represents a good approximation to the exact solution.

The integral (1.45) can be extended to the whole plane x-y by inserting a transmission function $\tau(x_o, y_o)$, which takes the value 1 inside the aperture Σ and 0 outside. The introduction of $\tau(x_o, y_o)$ allows for generalizing the treatment to partially transmitting screens and/or phase screens, in which case τ becomes a complex quantity with a modulus taking intermediate values between 0 and 1.

The generalized diffraction integral is written as:

$$E(\mathbf{r}) = \frac{i}{\lambda}\int_{-\infty}^{\infty} dx_o \int_{-\infty}^{\infty} \tau(x_o, y_o)E_i(\mathbf{r}_o)\frac{\exp(ik|\mathbf{r}-\mathbf{r}_o|)}{|\mathbf{r}-\mathbf{r}_o|}\cos\theta dy_o. \tag{1.48}$$

Diffraction theory, as expressed in (1.48), constitutes an approximate propagation theory of electromagnetic waves. In fact, once a field distribution is defined on the plane $z = 0$ immediately beyond the screen plane, the integral (1.48) gives the field distribution at any later generic plane z.

In many cases it is useful to consider the Fresnel approximation, which assumes that the angle θ is small for all the considered points. Mathematically, this is equivalent to the assumption that the fraction appearing inside the square root in the right-hand member of (1.47) is small compared to 1. By truncating the series expansion of (1.47) at the first order, the following expression is obtained:

$$|\mathbf{r}-\mathbf{r}_o| \approx z\left[1+\frac{(x-x_o)^2+(y-y_o)^2}{2z^2}\right]. \tag{1.49}$$

Considering the fraction inside the integral (1.48), the approximation $|\mathbf{r}-\mathbf{r}_o| = z$ can be used for the denominator, but not for the numerator. In general, in the case of an exponential of the type $\exp[i(A + A_1)]$, the assumption that A_1 is much smaller than A is not in itself a sufficient condition to neglect A_1. Since the exponential is a periodic function with period 2π, A_1 can be neglected only if it is much smaller than 2π.

By substituting (1.49) inside (1.48), and putting $\cos\theta = 1$, the diffraction integral under the Fresnel approximation becomes:

$$E(\mathbf{r}) = \frac{i\exp(ikz)}{\lambda z}\int_{-\infty}^{\infty} dx_o$$
$$\int_{-\infty}^{\infty} \tau(x_o, y_o)E_i(\mathbf{r}_o)\exp\left\{\frac{ik}{2z}[(x-x_o)^2+(y-y_o)^2]\right\} dy_o \qquad (1.50)$$

The integral (1.50) is a convolution of a complex Gaussian nucleus with the field distribution present in the plane immediately after the screen.

The Fresnel approximation is valid provided that the first neglected term at the exponent of the complex exponential is small compared to 2π:

$$\frac{k(x-x_o)^4}{4z^3} \ll 2\pi. \qquad (1.51)$$

Considering, as an example, a circular aperture of diameter D, (1.51) is verified if:

$$z^3 \gg \frac{D^4}{\lambda}. \qquad (1.52)$$

However, it should be mentioned that the comparison of the Fresnel approximation with exact numerical results shows a good agreement even in cases in which the condition (1.52) is not fully satisfied.

The Fresnel approximation is now applied to describe the free propagation of a Gaussian spherical wave. It is assumed that the wave has its waist on the plane $z = 0$. This means that the radius of curvature of the wavefront is infinite at $z = 0$, and thus: $1/q_o = i\lambda/(\pi w_o^2)$, where w_o is the beam radius at $z = 0$. The field distribution at $z = 0$ is:

$$E_i(x_o, y_o) = A_o\exp\left[i\frac{k(x_o^2+y_o^2)}{2q_o}\right]. \qquad (1.53)$$

The field distribution at the generic plane z is derived by inserting (1.53) into (1.50), and putting $\tau(x_o, y_o) = 1$. The diffraction integral is calculated by separating the integration variables. The result is:

$$E(x, y, z) = A_o\frac{i\exp(ikz)}{\lambda z}J(x, z)J(y, z), \qquad (1.54)$$

where:

$$J(x, z) = \int_{-\infty}^{\infty}\exp\left[i\frac{k(x-x_o)^2}{2z}\right]\exp\left[i\frac{kx_o^2}{2q_o}\right]dx_o. \qquad (1.55)$$

Recalling that:

$$\int_{-\infty}^{\infty}\exp\left(-\frac{x^2}{w^2}\right)dx = \int_{-\infty}^{\infty}\exp\left[-\frac{(x-a)^2}{w^2}\right]dx = \sqrt{\pi}\,w, \qquad (1.56)$$

where a is an arbitrary constant, the integral (1.55) is readily calculated by expressing the integrand function as a displaced Gaussian function:

$$J(x,z) = \int_{-\infty}^{\infty} \exp\left[\frac{ik}{2}\left(\frac{q_o+z}{q_o z}x_o^2 - \frac{2xx_o}{z} + \frac{x^2}{z}\right)\right] dx_o$$
$$= \exp\left[\frac{ikx^2}{2(q_o+z)}\right] \int_{\infty}^{\infty} \exp\left\{\frac{ik(q_o+z)}{2q_o z}\left[x_o - \frac{q_o x}{q_o+z}\right]^2\right\} dx_o. \quad (1.57)$$

The final expression of $E(x, y, z)$ is:

$$E(x,y,z) = A_o \frac{q_o}{q_o+z} e^{ikz} \exp\left[i\frac{k(x^2+y^2)}{2(q_o+z)}\right] \quad (1.58)$$

This result, which exactly matches (1.33), is not surprising because the Fresnel approximation of the diffraction integral is essentially equivalent to the paraxial approximation.

1.6 Fraunhofer Diffraction

A further simplification to diffraction theory is introduced by the Fraunhofer approximation, which consists of neglecting the terms quadratic in x_o and y_o in the exponent inside (1.50). This simplification is valid when the terms to be neglected are small in comparison to π. This requires:

$$z \gg \frac{(x_o^2+y_o^2)}{\lambda}. \quad (1.59)$$

For a circular aperture with diameter D, the maximum value of $x_o^2 + y_o^2$ is $D^2/4$, therefore the condition (1.59) becomes:

$$z \gg \frac{D^2}{4\lambda}. \quad (1.60)$$

In most practical cases it may be very difficult to satisfy such a condition. If, for instance, $\lambda = 0.6\,\mu\text{m}$, and $D = 5\,\text{cm}$, then (1.60) requires $z \gg 3\,\text{km}$. This example explains why the Fraunhofer approximation is also called the far-field approximation. As it will be described in the next chapter, a converging lens can be used to bring the far field into the lens focal plane. By exploiting this possibility, the Fraunhofer approximation can be utilized in various applications.

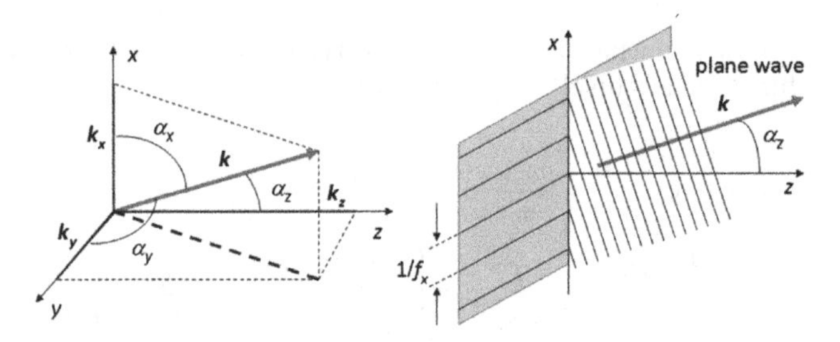

Fig. 1.8 Decomposition of an optical wave into plane waves

The integral (1.50), as written in the Fraunhofer approximation, becomes:

$$E(x,y,z)=\frac{i\exp(ikz)}{\lambda z}\exp\left[\frac{ik}{2z}(x^2+y^2)\right]$$
$$\int_{-\infty}^{\infty}dx_o\int_{-\infty}^{\infty}\tau(x_o,y_o)E_i(x_o,y_o)\exp[-i(k/z)(xx_o+yy_o)]dy_o. \quad (1.61)$$

Apart from a multiplicative factor, the field distribution $E(x,y,z)$ given by (1.61) is the two-dimensional Fourier transform of the distribution at $z=0$. The two variables that are conjugated to x_o, y_o are: $f_x=x/(\lambda z)$ and $f_y=y/(\lambda z)$. These are the components of the spatial frequency **f**, having modulus $f=\sqrt{f_x^2+f_y^2}$.

In order to grasp the physical meaning of the quantity **f**, it should be recalled that the field τE_i present immediately after the plane x_o-y_o can always be described as a superposition of plane waves. As shown in Fig. 1.8, each plane wave is characterized by a different vector **k**. The direction of **k** is determined by assigning the two angles α_x and α_y, formed by **k** with the axes x and y. Once given α_x and α_y, the angle α_z between **k** and z may be calculated from the relation: $\cos^2\alpha_x+\cos^2\alpha_y+\cos^2\alpha_z=1$. For every pair of values of α_x and α_y, there corresponds a spatial frequency **f** with components $f_x=\lambda^{-1}\cos\alpha_x$ e $f_y=\lambda^{-1}\cos\alpha_y$. The modulus of **f** is:

$$f=\frac{\sqrt{1-\cos^2\alpha_z}}{\lambda}=\frac{\sin\alpha_z}{\lambda}. \quad (1.62)$$

The term "spatial frequency" comes from the fact that f^{-1} is the spacing (spatial period) of the projected wavefronts on the plane x_o-y_o, as shown in Fig. 1.8. If the plane wave propagates perpendicularly to z, $\alpha_z=\pi/2$ and $f^{-1}=\lambda$. If **k** is parallel to z, $\alpha_z=0$, and f^{-1} becomes infinite.

Equation (1.62) indicates that the plane wave characterized by α_x and α_y is represented on the plane z in the far field by the point $x=f_x\lambda z$ and $y=f_y\lambda z$. In the case where the field incident on the screen is a plane wave directed along z, $E(x,y,z)$ is simply the Fourier transform of the transmission function $\tau(x_o,y_o)$.

1.6.1 Rectangular and Circular Apertures

As a first example of the application of the Fraunhofer approximation, a rectangular aperture illuminated by a plane wave propagating along z is here considered. The assumed transmission function is:

$$\tau(x_o, y_o) = rect\left(\frac{x_o}{L_x}\right) rect\left(\frac{y_o}{L_y}\right), \tag{1.63}$$

where $rect(x)$ is unitary inside the interval $-0.5 \leq x \leq 0.5$ and is 0 elsewhere. The transform of (1.63) is: $F\{\tau(x_o, y_o)\} = L_x L_y \sin c(\pi L_x f_x) \sin c(\pi L_y f_y)$ where $\mathrm{sinc}(x) = \sin x/x$. By substitution into (1.61), it is found:

$$E(x, y, z) = A_o \frac{i \exp(ikz)}{\lambda z} \exp\left[\frac{ik}{2z}(x^2 + y^2)\right] L_x L_y \,\mathrm{sinc}\left(\pi \frac{x L_x}{\lambda z}\right) \mathrm{sinc}\left(\pi \frac{y L_y}{\lambda z}\right), \tag{1.64}$$

where A_o is the amplitude of the incident wave.

Figure 1.9 shows the image of the diffracted wave, whereas Fig. 1.10 illustrates the transversal behavior of the optical intensity $I(x, y, z)$, proportional to $|E(x, y, z)|^2$, along the x axis. While the wave incident on the aperture consists only of the propagation vector parallel to z, the diffracted wave has a distribution in **k** vectors that becomes wider as the size of the aperture reduces. In fact, the width of the main diffraction lobe, evaluated along x as the distance between the first two zeroes, is: $\Delta x = \lambda z/L_x$, or, in other terms, the angular aperture of the main diffraction lobe is $\lambda/(2L_x)$.

As a second example, consider a circular aperture having a transmission function:

$$\tau(x_o, y_o) = \mathrm{circ}\left(\frac{r_o}{R}\right), \tag{1.65}$$

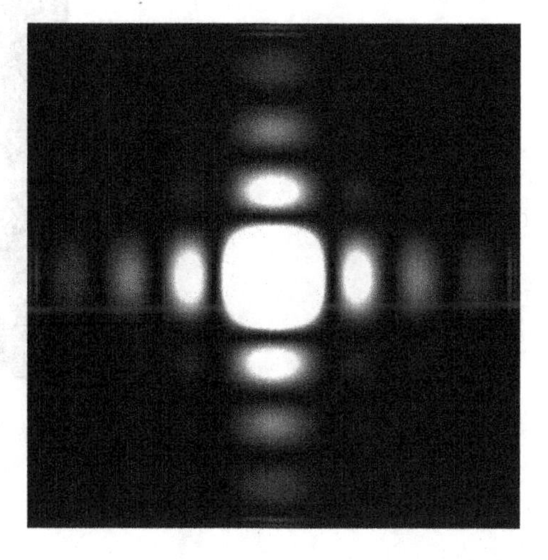

Fig. 1.9 Fraunhofer diffraction from a square aperture: image of the diffracted wave

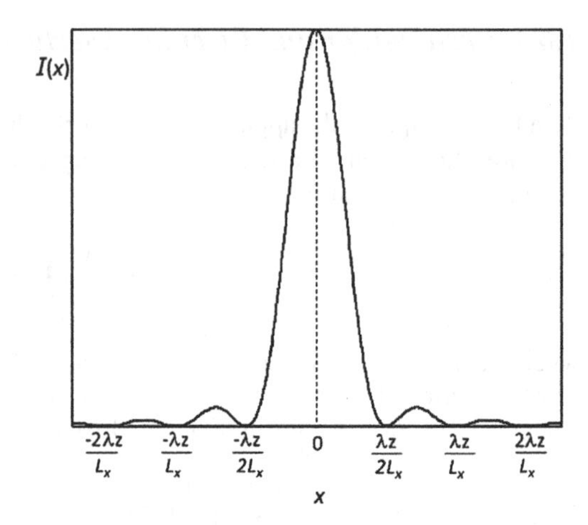

Fig. 1.10 Fraunhofer diffraction from a square aperture: intensity versus x

where $\text{circ}(r_o/R)$ is equal to 1 inside the circle of radius R and 0 elsewhere. Using (1.61), the diffracted field is found to be:

$$E(r) = A_o \frac{i\exp(ikz)}{\lambda z} \exp\left[\frac{ik}{2z}r^2\right] R^2 \frac{J_1[rR/(\lambda z)]}{rR/(\lambda z)}, \tag{1.66}$$

where J_1 is the order 1 Bessel function. The intensity distribution $I(r)$, proportional to $|E(r)|^2$, is called Airy function. Figure 1.11 shows the image of the diffracted wave, and Fig. 1.12 gives the behavior of $I(r)$. The radius corresponding to the first zero is: $0.61\,\lambda z/R$. Here again it is found that the angular spread of the diffracted wave is of the order of the ratio between wavelength and size of the aperture.

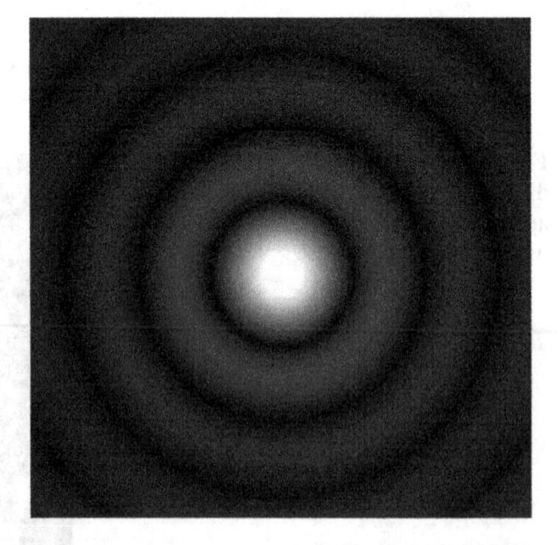

Fig. 1.11 Fraunhofer diffraction from a circular aperture: image of the diffracted wave

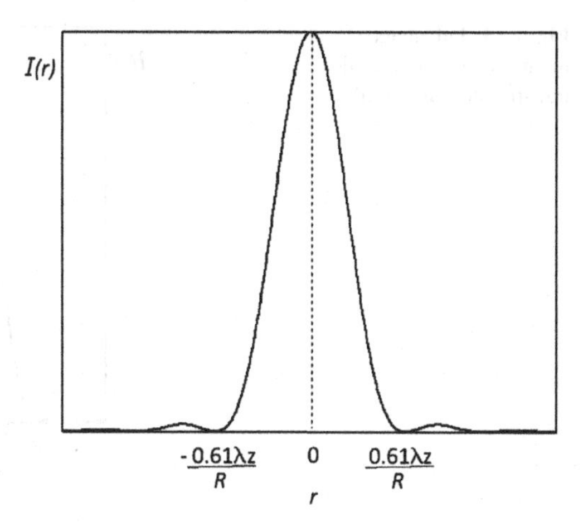

Fig. 1.12 Fraunhofer diffraction from a circular aperture: behavior of the diffracted intensity $I(r)$ as a function of the distance from the z axis

1.6.2 Periodical Transmission Function

Consider a plane wave propagating along z that illuminates a square aperture of side L that has a transmission function varying sinusoidally along x, with period $d = 1/f_o$:

$$\tau(x_o, y_o) = \frac{1 + m\cos(2\pi f_o x_o)}{2} rect\left(\frac{x_o}{L}\right) rect\left(\frac{y_o}{L}\right), \tag{1.67}$$

where the parameter m is the modulation depth.

The Fourier transform of the fraction that appears in the right-hand member of (1.67) is:

$$F\left\{\frac{1 + m\cos 2\pi f_o x_o}{2}\right\} = \frac{1}{2}\delta(f_x, f_y) + \frac{m}{4}\delta(f_x + f_o, f_y) + \frac{m}{4}\delta(f_x - f_o, f_y), \tag{1.68}$$

where $\delta(x)$ is the Dirac delta function, which is infinite at $x = 0$ and zero elsewhere. Recalling that the Fourier transform of the product of two functions is the convolution of the two transforms, the far-field diffraction pattern is given by:

$$E(x, y, z) = A_o \frac{iL^2 \exp(ikz)}{2\lambda z} \exp\left[\frac{ik}{2z}(x^2 + y^2)\right] \text{sinc}\left(\pi \frac{yL}{\lambda z}\right)$$
$$\left\{\text{sinc}\left(\pi \frac{xL}{\lambda z}\right) + \frac{m}{2}\,\text{sinc}\left[\pi \frac{L}{\lambda z}(x + f_o \lambda z)\right] + \frac{m}{2}\,\text{sinc}\left[\pi \frac{L}{\lambda z}(x - f_o \lambda z)\right]\right\}. \tag{1.69}$$

The behavior of the diffracted intensity as a function of x is reported in Fig. 1.13. The three peaks correspond to three diffracted waves. The central peak, i.e., the zero-th order wave, represents the fraction of incident wave that is still propagating along

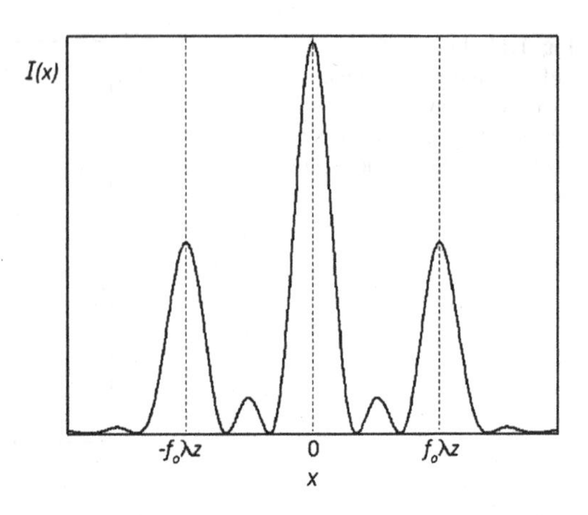

Fig. 1.13 Diffracted intensity from a periodical transmission function

z, with a distribution of **k** vectors arising from the diffracting effect due to the finite size of the aperture. The lateral peaks, symmetrically positioned around the central one, represent waves with propagation vectors laying in the x-z plane and forming the angles $\pm\lambda/d$ with the z axis. These two waves constitute the diffraction order $+1$ and -1, respectively. The spatial separation between order 0 and order ± 1 is $f_o\lambda z$, which means that the angular separation is $\pm f_o\lambda$. The two first-order diffraction peaks are broadened, like the central peak, because of the finite size of the aperture.

At this point it is easy to understand what happens if the transmission function $\tau(x_o, y_o)$ is a periodic function with a square profile instead of a sinusoidal profile. The square profile contains, with decreasing amplitude, all the harmonics. As a consequence, diffracted waves with orders higher than 1 are generated.

An other interesting case is that of a phase screen, having a transmission function that is a complex function with unitary modulus and sinusoidal phase:

$$\tau(x_o, y_o) = \exp\left[i\frac{m\sin(2\pi f_o x_o)}{2}\right] rect\left(\frac{x_o}{L}\right) rect\left(\frac{y_o}{L}\right). \tag{1.70}$$

Such a transfer function can be realized by using a transparent material with a periodic modulation of index of refraction or of thickness along the coordinate x_o. Recalling that:

$$\exp\left[i\frac{m\sin 2\pi f_o x_o}{2}\right] = \sum_{q=-\infty}^{\infty} J_q\left(\frac{m}{2}\right)\exp(i2\pi q f_o x_o), \tag{1.71}$$

where J_q is the q-th order first-type Bessel function, and using the convolution theorem, one finds:

$$E(x, y, z) = A_o \frac{iL^2 \exp(ikz)}{2\lambda z} \exp\Big[\frac{ik}{2z}(x^2 + y^2)\Big] \sin c\Big(\pi \frac{yL}{\lambda z}\Big)$$
$$\sum_{q=-\infty}^{\infty} J_q\left(\frac{m}{2}\right) \sin c\left[\frac{\pi L}{\lambda z}(x - q f_o \lambda z)\right] \qquad (1.72)$$

Equation (1.72) shows that the sinusoidal phase profile presents, in principle, all possible diffraction orders. The amplitude of the generic order q is determined by the value of the corresponding Bessel function for the argument $m/2$. Note that the transmitted beam (that is, the zero-th order) can be totally suppressed in the particular case of $m/2$ coinciding with a zero of J_0.

As it will be seen in the next chapter, the diffraction from periodical transmission profiles has important applications to spectral measurements performed by using gratings. The Fraunhofer approximation allows calculating not only the geometrical properties, but also the power partitioning over the various diffraction orders of the grating.

Problems

1.1 Consider the plane electromagnetic wave whose electric field is given by the expressions (in SI units) $E_x = 0$, $E_y = 200 \cos[12\pi \times 10^{14}(t - z/c) + \pi/2]$, and $E_z = 0$. What are the frequency, wavelength, direction of motion, amplitude, initial phase angle, and polarization of the wave?

1.2 Consider the plane electromagnetic wave in vacuum whose electric field is given by the expression $E = E_o \cos(\omega t - kz)$. Assuming that the electric field amplitude E_o is 100 V/m, calculate the amplitude of the magnetic field B_o in units of Tesla and the intensity I in units of W/m^2.

1.3 A plane electromagnetic wave propagating along the z axis has the following electric field components: $E_x(t, z) = E_o \cos(\omega t - kz)$, and $E_y(t, z) = \sqrt{3} E_o \cos(\omega t - kz)$. Determine the angle formed by its polarization direction with the x axis.

1.4 A Gaussian beam at wavelength $\lambda = 0.63\,\mu$m has a radius at the beam waist of $w_o = 0.5$ mm. Calculate the beam radius w and the radius of curvature of the wavefront at a distance of 5 m from the beam waist.

1.5 A plane electromagnetic wave has wavelength $\lambda = 500$ nm and a wave vector **k** that lies in the x-z plane, forming an angle $\alpha_x = 60°$ with the x axis. Determine the spatial frequency of the phase pattern on the $z = 0$ plane.

1.6 A plane wave of wavelength $\lambda = 500$ nm illuminates a circular aperture having diameter $D = 100\,\mu$m. The diffraction pattern is observed on a screen placed at a distance of 30 cm from the plane of the aperture. (i) check whether the condition for Fraunhofer diffraction is satisfied; (ii) determine the position of the first zero of the Airy function.

Chapter 2
Optical Components and Methods

Abstract In this chapter, after presenting Maxwell equations inside matter and describing reflection and refraction processes, the interaction of optical waves with simple optical components such as mirrors, prisms, and lenses is examined. The different methods for measuring the power spectrum of an optical beam are described and compared in Sect. 2.3. In Sect. 2.4, after treating wave propagation in anisotropic materials, the methods used for fixing or modifying the polarization state of an optical beam are presented. Finally, Sect. 2.5 introduces the important subject of optical waveguides.

2.1 Electromagnetic Waves in Matter

Inside a medium Maxwell's equations take the following form:

$$\nabla \cdot \mathbf{D} = \rho \tag{2.1}$$

$$\nabla \cdot \mathbf{B} = 0 \tag{2.2}$$

$$\nabla \times \mathbf{H} = \mathbf{J} + \frac{\partial \mathbf{D}}{\partial t} \tag{2.3}$$

$$\nabla \times \mathbf{E} = -\frac{\partial \mathbf{B}}{\partial t}, \tag{2.4}$$

where **D** is the electric displacement vector (measured in units of C/m^2). The quantities ρ and **J** are, respectively, the electric charge density (C/m^3) and the current density (A/m^2). Here the treatment is limited to dielectric media, in which there are no free electric charges or currents, so that one can put $\mathbf{J} = \rho = 0$.

The relations connecting **D** to **E** and **B** to **H** inside the medium become:

$$\mathbf{D} = \varepsilon_o \mathbf{E} + \mathbf{P} \tag{2.5}$$

$$\mathbf{B} = \mu_o (\mathbf{H} + \mathbf{M}). \tag{2.6}$$

V. Degiorgio and I. Cristiani, *Photonics*, Undergraduate Lecture Notes in Physics,
DOI 10.1007/978-3-319-20627-1_2

The new vectors **P** and **M** are, respectively, the unit-volume electric polarization and the unit-volume magnetization inside the medium. In this book non-ferromagnetic media are only considered, and so it is possible to simplify the treatment by putting $\mathbf{M} = 0$. The wave equation inside the medium is the following:

$$\nabla^2 \mathbf{E} = \frac{1}{c^2} \frac{\partial^2 \mathbf{E}}{\partial t^2} + \mu_o \frac{\partial^2 \mathbf{P}}{\partial t^2}. \tag{2.7}$$

In order to solve wave propagation problems, it is necessary to associate an equation relating **P** to **E** to (2.7). It is useful to recall what is the physical meaning of **P**. The medium is a collection of positively charged nuclei and negatively-charged bound electrons. Under the action of an electric field the electron clouds are slightly displaced from their equilibrium position, so that a macroscopic electric dipole is created. Limiting the discussion to the time-dependence, it is clear that the medium response to the application of an electric field cannot be instantaneous. Therefore, a convolution relation should be used:

$$\mathbf{P}(\mathbf{r}, t) = \varepsilon_o \int_{-\infty}^{t} R(t - t')\mathbf{E}(\mathbf{r}, t')dt', \tag{2.8}$$

where $R(t - t')$ is the response function of the medium.

The electric field $E(\mathbf{r}, t)$ associated with a generic light beam can always be described as a weighted superposition of sinusoidal functions oscillating at different frequencies, as expressed by the integral:

$$E(\mathbf{r}, t) = \frac{1}{2\pi} \int_{-\infty}^{\infty} d\omega E(\mathbf{r}, \omega) \exp(-i\omega t). \tag{2.9}$$

The complex quantity $E(\mathbf{r}, \omega)$ represents the weight of the component at angular frequency ω. From a mathematical point of view, $E(\mathbf{r}, \omega)$ is the Fourier transform of $E(\mathbf{r}, t)$.

The inversion of (2.9) gives:

$$E(\mathbf{r}, \omega) = \int_{-\infty}^{\infty} dt E(\mathbf{r}, t) \exp(i\omega t). \tag{2.10}$$

Recalling that the Fourier transform of the convolution of two functions is the product of their Fourier transforms, the Fourier transform of $\mathbf{P}(\mathbf{r}, t)$ is expressed by:

$$\mathbf{P}(\mathbf{r}, \omega) = \varepsilon_o \chi(\omega) \mathbf{E}(\mathbf{r}, \omega), \tag{2.11}$$

where $\chi(\omega)$, called electric susceptibility, is given by:

$$\chi(\omega) = \int_{-\infty}^{\infty} R(t) \exp(i\omega t) dt. \tag{2.12}$$

By substituting (2.11) inside (2.5), it is found that the electric displacement $\mathbf{D}(\mathbf{r}, \omega)$, generated by a monochromatic electric field oscillating at the frequency ω, is proportional to $\mathbf{E}(\mathbf{r}, \omega)$:

$$\mathbf{D}(\mathbf{r}, \omega) = \varepsilon_o(1 + \chi)\mathbf{E}(\mathbf{r}, \omega) = \varepsilon_o\varepsilon_r\mathbf{E}(\mathbf{r}, \omega). \tag{2.13}$$

The dimensionless constant ε_r is called relative dielectric constant or relative electric permittivity. The wave propagation velocity inside the dielectric medium is:

$$u = \frac{1}{\sqrt{\varepsilon_o\varepsilon_r\mu_o}} = \frac{c}{n} \tag{2.14}$$

The quantity n, called index of refraction of the medium, is given by:

$$n = \sqrt{\varepsilon_r} = \sqrt{1 + \chi}. \tag{2.15}$$

In the more general case, in which the relative magnetic permeability of the material, μ_r, may be different from 1, it is found that $n = \sqrt{\varepsilon_r\mu_r}$.

Since the frequency of oscillation cannot change when the wave goes from one medium to the other, the change of velocity is associated with a change of wavelength. Specifically, if λ is the vacuum wavelength, then the wavelength inside a medium of index of refraction n is: $\lambda' = \lambda/n$.

Since $\chi(\omega)$ is frequency-dependent, also $n(\omega) = \sqrt{1 + \chi(\omega)}$ is frequency-dependent. This phenomenon is called optical dispersion. As an example, over its transparency window going from 0.35 to 2.3 μm, the index of refraction of the borosilicate glass called BK7 decreases monotonically from 1.54 at 0.35 μm to 1.49 at 2.3 μm. This behavior is called normal dispersion. The dispersion is said to be anomalous when the index of refraction increases as a function of wavelength. As it will be seen in Chap. 6, the propagation of ultrashort light pulses is strongly influenced by optical dispersion.

The electric field of the monochromatic plane wave inside the medium is still given by (1.9). The only difference is that the modulus of the propagation vector is now:

$$k = \frac{\omega}{u} = \frac{\omega}{c}\sqrt{1 + \chi} = \frac{\omega}{c}n, \tag{2.16}$$

and (1.12) becomes:

$$I = c\frac{\varepsilon_o n E_o^2}{2}. \tag{2.17}$$

The propagation inside a medium with a real χ proceeds similarly to vacuum propagation, the only difference being in the propagation velocity. However, if χ is a complex quantity, then n is also complex, and can be written as: $n = n' + in''$. Assuming that a monochromatic plane wave, traveling along z, enters into the medium at $z = 0$, the electric field at the generic coordinate z is:

$$\mathbf{E} = \mathbf{E_o}\exp\{-i([\omega t - (\omega/c)(n' + in'')z]\}. \tag{2.18}$$

The intensity $I(z)$, which is proportional to the square of the field modulus, is given by:

$$I(z) = I_o\exp[-2(\omega/c)n''z] = I_o\exp(-\alpha z), \tag{2.19}$$

where I_o is the intensity at $z = 0$. Equation (2.19) shows that, if $n'' > 0$, the wave intensity decays exponentially during propagation with an attenuation coefficient, $\alpha = 2(\omega/c)n''$, proportional to the imaginary part of the refractive index. The attenuation may be due to absorption or to scattering.

2.2 Reflection and Refraction

In this section the laws of reflection and refraction are introduced and applied to describe the behavior of different optical components, such as mirrors and lenses.

2.2.1 *Dielectric Interface*

A monochromatic plane wave is incident at the planar boundary between two dielectric media, characterized by refractive indices n_1 and n_2. According to the geometry shown in Fig. 2.1, the wave is coming from medium 1, and its wave-vector $\mathbf{k_i}$ forms an angle θ_i with the normal to the boundary. The planar boundary is taken to coincide with the x-y plane, and $\mathbf{k_i}$ lies on the y-z plane. The plane containing $\mathbf{k_i}$ and the normal to the planar boundary is known as the incidence plane. In the case of Fig. 2.1, y-z is the incidence plane.

Part of the wave will be reflected and part will be transmitted. Let θ_r and θ_t be the reflection and transmission angles, respectively. By imposing the continuity conditions at the boundary for the tangential components of the electric and magnetic

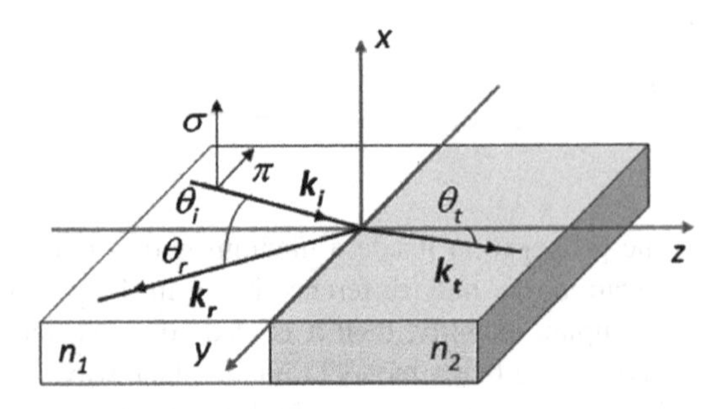

Fig. 2.1 Reflection and refraction at a dielectric boundary

fields, it is possible to calculate the direction and the complex amplitude of the reflected and transmitted (also called refracted) waves.

It is useful to consider separately two different polarization states of the incident wave: (a) a wave linearly polarized in a direction perpendicular to the incidence plane, this is usually called σ case; (b) a wave linearly polarized in the incidence plane, this is usually called π case. Any arbitrary polarization state of the incident wave can always be described as a linear superposition of these two states.

• σ case.

Putting $|\mathbf{k_i}| = k_o n_1$, where $k_o = 2\pi/\lambda$, the electric field of the incident wave is:

$$\mathbf{E_i} = E_i \exp(-i\omega t + ik_o y n_1 \sin\theta_i + ik_o z n_1 \cos\theta_i)\mathbf{x}, \tag{2.20}$$

where $\mathbf{x}$ is the unit vector directed along x. The electric fields $\mathbf{E_r}$ and $\mathbf{E_t}$ of the reflected and transmitted waves are written as:

$$\mathbf{E_r} = E_r \exp(-i\omega t + ik_o y n_1 \sin\theta_r - ik_o z n_1 \cos\theta_r)\mathbf{x}, \tag{2.21}$$

and

$$\mathbf{E_t} = E_t \exp(-i\omega t + ik_o y n_2 \sin\theta_t + ik_o z n_2 \cos\theta_t)\mathbf{x}. \tag{2.22}$$

The continuity condition for the tangential component of the electric field at $z = 0$ is:

$$E_i \exp(ik_o y n_1 \sin\theta_i)\mathbf{x} + E_r \exp(ik_o y n_1 \sin\theta_r)\mathbf{x} = E_t \exp(ik_o y n_2 \sin\theta_t)\mathbf{x}. \tag{2.23}$$

In order to satisfy (2.23) for any y, the reflection and transmission angles should satisfy the relations:

$$\theta_i = \theta_r, \tag{2.24}$$

and

$$n_1 \sin\theta_i = n_2 \sin\theta_t. \tag{2.25}$$

Equation (2.25) is known as Snell's law.

By using (2.24) and (2.25), (2.23) simply becomes

$$E_i + E_r = E_t \tag{2.26}$$

In order to derive E_r and E_t, a second equation is needed besides that of (2.26). The continuity of the tangential component of the magnetic field is expressed as:

$$H_i \cos\theta_i - H_r \cos\theta_r = H_t \cos\theta_t \tag{2.27}$$

Recalling that the electric field amplitude is related to the magnetic field amplitude by the expression $H = \sqrt{\varepsilon_r \varepsilon_o/\mu_o} E$, (2.27) can be written as:

$$E_i \cos\theta_i - E_r \cos\theta_r = \frac{n_2}{n_1} E_t \cos\theta_t. \tag{2.28}$$

The solution of the system of (2.26) and (2.28) can be presented in terms of the transmission and reflection coefficients, τ_σ and ρ_σ, respectively:

$$\tau_\sigma = \frac{E_t}{E_i} = \frac{2n_1 \cos\theta_i}{n_1 \cos\theta_i + n_2 \cos\theta_t} = \frac{2 \sin\theta_t \cos\theta_i}{\sin(\theta_i + \theta_t)}, \tag{2.29}$$

$$\rho_\sigma = \frac{E_r}{E_i} = \frac{n_1 \cos\theta_i - n_2 \cos\theta_t}{n_1 \cos\theta_i + n_2 \cos\theta_t} = -\frac{\sin(\theta_i - \theta_t)}{\sin(\theta_i + \theta_t)}. \tag{2.30}$$

If the incident beam is coming from medium 2, one can repeat the treatment finding two new coefficients, τ'_σ and ρ'_σ, related to the previous ones as follows:

$$\rho'_\sigma = -\rho_\sigma; \quad \tau_\sigma \tau'_\sigma = 1 - \rho_\sigma^2 \tag{2.31}$$

- π case.

Following a treatment similar to that of the σ case, it is found that:

$$\tau_\pi = \frac{E_t}{E_i} = \frac{2n_1 \cos\theta_i}{n_1 \cos\theta_t + n_2 \cos\theta_i} = \frac{2 \sin\theta_t \cos\theta_i}{\sin(\theta_i + \theta_t) \cos(\theta_i - \theta_t)}, \tag{2.32}$$

$$\rho_\pi = \frac{E_r}{E_i} = \frac{n_1 \cos\theta_t - n_2 \cos\theta_i}{n_1 \cos\theta_t + n_2 \cos\theta_i} = -\frac{\tan(\theta_i - \theta_t)}{\tan(\theta_i + \theta_t)}. \tag{2.33}$$

If the beam arrives at the boundary from medium 2, it is found that the relations between the coefficients τ'_π, ρ'_π and the coefficients τ_π and ρ_π are identical to those of (2.31).

In the case of normal incidence, $\theta_i = \theta_r = \theta_t = 0$, and ρ_σ coincides with ρ_π:

$$\rho_\sigma(\theta_i = 0) = \rho_\pi(\theta_i = 0) = -\frac{n_2 - n_1}{n_2 + n_1}. \tag{2.34}$$

If $n_2 > n_1$, the field reflection coefficient is negative. This indicates that the phase of E_r is changed by π with respect to E_i. If $n_2 < n_1$, E_r has the same phase as E_i.

In order to derive how the incident power splits between transmission and reflection, it should be recalled that the power arriving on a generic surface Σ, is determined as the time average of the flux of the Poynting vector through the surface. Projecting the Poynting vector on the direction perpendicular to the surface, allows the following expression for P_i to be given:

$$P_i = (1/2) c \varepsilon_o n_1 \Sigma |E_i|^2 \cos\theta_i. \tag{2.35}$$

Similarly, for the transmitted and reflected power:

$$P_r = (1/2)c\varepsilon_o n_1 \Sigma |E_r|^2 \cos\theta_i \qquad P_t = (1/2)c\varepsilon_o n_2 \Sigma |E_t|^2 \cos\theta_t. \tag{2.36}$$

Therefore the reflectance R and the transmittance T are given by:

$$R = \frac{P_r}{P_i} = \frac{|E_r|^2}{|E_i|^2} = |\rho|^2 \tag{2.37}$$

$$T = \frac{P_t}{P_i} = |\tau|^2 \frac{n_2 \cos\theta_t}{n_1 \cos\theta_i}. \tag{2.38}$$

Note that $R + T = 1$, whereas the sum $|\rho|^2 + |\tau|^2$ is, in general, $\neq 1$.

The behavior of R as a function of θ_i depends on whether n_1 is smaller or larger than n_2. Therefore the two cases are discussed separately.

- $n_1 < n_2$

Considering, as an example, an air/glass boundary with a ratio $n_2/n_1 = 1.5$, the behavior of R_π e R_σ for a wave coming from air is illustrated in Fig. 2.2. At normal incidence ($\theta_i = 0$), using (2.34), it is found that $R_\pi(\theta_i = 0) = R_\sigma(\theta_i = 0) = 0.04$. The reflectance R_σ is monotonically increasing with θ_i reaching a value of 1 at $\theta_i = \pi/2 = 90°$. On the other hand, R_π at first decreases till it vanishes at $\theta_i = \theta_B$, and then increases, becoming equal to 1 at $\theta_i = \pi/2$.

The angle θ_B is derived by noting that the reflection coefficient ρ_π vanishes if the denominator of the fraction at right-hand side of (2.33) becomes infinite. This happens if: $\theta_i + \theta_t = \pi/2$. From this condition, by using Snell's law, the following expression of θ_B, called Brewster's angle, is derived:

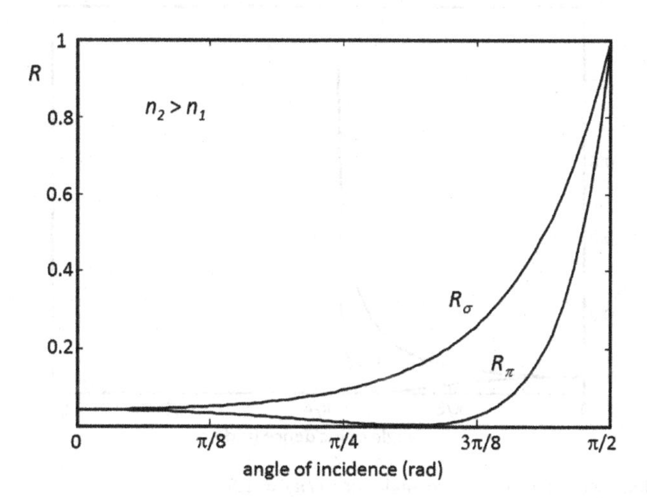

Fig. 2.2 Reflectance versus incidence angle for $n_2/n_1 = 1.5$

$$\tan\theta_B = \frac{n_2}{n_1} \tag{2.39}$$

As an example, if $n_2/n_1 = 1.5$, $\theta_B = 56°$, and $R_\sigma(\theta_B) = 0.147$.

In general, ρ is a complex quantity that can be written as $\rho = \sqrt{R}\exp(i\phi)$, where ϕ is the phase shift of the reflected field with respect to the incident field. In the σ case, (2.30) shows that $\phi_\sigma = \pi$ over the whole interval $0 \le \theta_i \le \pi/2$. In the π case, (2.33) indicates that ϕ_π is equal to π within the interval $0 \le \theta_i < \theta_B$, and vanishes for $\theta_B \le \theta_i \le \pi/2$.

- $n_1 > n_2$

In this case, Snell's law indicates that θ_t is larger than θ_i. Therefore, as θ_i grows, there will be an incidence angle, θ_c, smaller than $\pi/2$, at which θ_t becomes $\pi/2$. If $\theta_i \ge \theta_c$ there is no transmitted beam, and so the power reflection coefficient is equal to 1. The total reflection angle or limit angle, θ_c, is defined by the relation:

$$\sin\theta_c = n_2/n_1 \tag{2.40}$$

Considering the glass/air interface, $n_1/n_2 = 1.5$, for which $\theta_c = 41.8°$, the behavior of R_π and R_σ is illustrated in Fig. 2.3. As in the previous case, R_π vanishes at the Brewster angle, $\theta_B = 33.7°$.

The field reflection coefficient is a real quantity for $\theta_i \le \theta_c$. If $\theta_i > \theta_c$, $\cos\theta_t$ becomes imaginary:

$$\cos\theta_t = \sqrt{1-\sin^2\theta_t} = \sqrt{1-\frac{n_1^2}{n_2^2}\sin^2\theta_i} = i\sqrt{\frac{n_1^2}{n_2^2}\sin^2\theta_i - 1}. \tag{2.41}$$

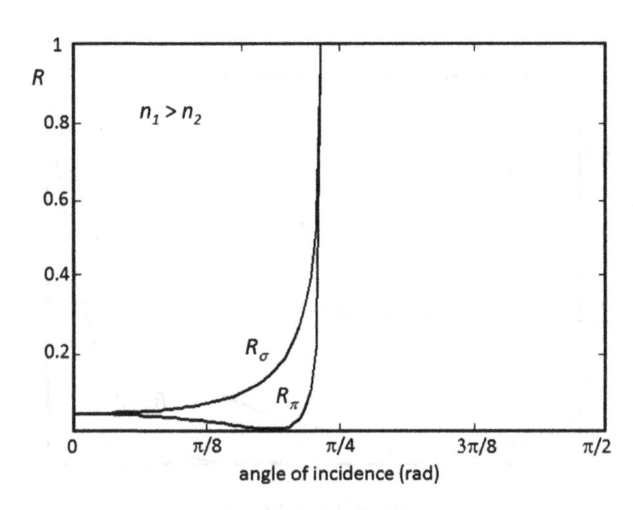

Fig. 2.3 Reflectance versus incidence angle for $n_1/n_2 = 1.5$

By substituting (2.41) into the expressions for ρ, it is seen that the field reflection coefficient becomes complex. In the σ case, the phase ϕ_σ of ρ_σ is 0 for $\theta_i \le \theta_c$, and grows from 0 to π in the interval $\theta_c \le \theta_i \le \pi/2$ according to the expression:

$$\tan\left(\frac{\phi_\sigma}{2}\right) = \frac{\sqrt{\sin^2\theta_i - \sin^2\theta_c}}{\cos\theta_i}. \tag{2.42}$$

In the π case, ϕ_π vanishes in the interval $0 \le \theta_i \le \theta_B$, and is equal to π in the interval $\theta_B < \theta_i \le \theta_c$, decreasing from π to 0 in the interval $\theta_c \le \theta_i \le \pi/2$ according to the expression:

$$\tan\left(\frac{\phi_\pi}{2}\right) = \frac{\cos\theta_i \sin^2\theta_c}{\sqrt{\sin^2\theta_i - \sin^2\theta_c}}. \tag{2.43}$$

2.2.2 Reflection from a Metallic Surface

The most commonly used mirrors are made by depositing a thin metallic layer on a glass substrate. It is therefore interesting to discuss the case in which medium 2 in Fig. 2.1 is a metal. Metals contain electrons that are free to move inside the crystal. In a simplified approach, the relative dielectric constant of the metal is written as the sum of two terms, one due to bound electrons, ε', and the other related to free electrons:

$$\varepsilon_m = \varepsilon' + i\frac{\sigma}{\varepsilon_o\omega}, \tag{2.44}$$

where σ is the frequency-dependent electric conductivity of the metal:

$$\sigma(\omega) = \sigma(0)\frac{1 + i\omega\tau}{1 + \omega^2\tau^2}. \tag{2.45}$$

The quantity τ is a relaxation time that depends on the nature of the metal. The zero-frequency conductivity is given by:

$$\sigma(0) = \frac{Ne^2\tau}{m_e}, \tag{2.46}$$

where N is the density of free electrons, m_e is the electron mass, and e is the electron charge.

The free electron gas inside the metal has a characteristic frequency of oscillation, called the plasma frequency, related to $\sigma(0)$ by the expression:

$$\omega_p = \sqrt{\frac{\sigma(0)}{\varepsilon' \varepsilon_o \tau}} = \sqrt{\frac{Ne^2}{\varepsilon' \varepsilon_o m_e}} \tag{2.47}$$

Electromagnetic radiation having frequency close to the plasma frequency is strongly absorbed by the metal, and so the refractive index must be written as a complex quantity, $n_m = \sqrt{\varepsilon_m} = n' + in''$. Limiting the discussion to normal incidence ($\theta_i = 0$), the reflectance of the air-metal interface, derived by using (2.34), is given by:

$$R = \frac{|n_m - 1|^2}{|n_m + 1|^2} = \frac{n'^2 + n''^2 + 1 - 2n'}{n'^2 + n''^2 + 1 + 2n'}. \tag{2.48}$$

Approximate expressions for n' and n'' can be derived from (2.44) by assuming that the contribution of bound electrons is negligible in comparison to that of free electrons. Taking into account that, for a typical metal, $\tau \approx 10^{-13}$ s and $\omega_p \approx 10^{15}\ \mathrm{s}^{-1}$, three different frequency intervals can be distinguished:

- infrared: $\omega \ll 1/\tau \ll \omega_p$. The result is: $R = 1$.
- near infrared and visible: $1/\tau \ll \omega \ll \omega_p$. It is found that: $R = 1 - 2/(\omega_p \tau)$, the reflectance is large, but lower than 100 %.
- ultraviolet: $\omega_p \ll \omega$. It is found: $R = \omega_p^4/(16\omega^6\tau^2) \ll 1$. In this case the reflectance is very low.

Experimental values of n', n'' and R for three metals at three different wavelengths are reported in Table 2.1. The data refer to thin layers deposited by evaporation on a glass substrate.

Table 2.1 Complex index of refraction and normal-incidence reflectance for some metal surfaces at three distinct wavelengths

Metal	λ (nm)	n′	n″	R
Silver	400	0.08	1.9	0.94
	550	0.06	3.3	0.98
	700	0.08	4.6	0.99
Aluminum	400	0.4	4.5	0.93
	550	0.8	6	0.92
	700	1.5	7	0.89
Gold	450	1.4	1.9	0.40
	550	0.33	2.3	0.82
	700	0.13	3.8	0.97

One problem with metallic mirrors is that the fraction of incident power that is not reflected is absorbed. When reflecting high-intensity beams, such as those coming from laser sources, the temperature rise due to this absorption may cause damage to the metallic layer.

2.2.3 Anti-reflection Coating

The reflectance of a boundary separating two transparent media can be strongly modified by inserting thin dielectric layers. The structure investigated in this section, shown in Fig. 2.4, consists of a single layer of refractive index n_2 and thickness d separating media 1 and 3 which have refractive indices n_1 and n_3, respectively. It is assumed that $n_1 < n_2 < n_3$.

Considering a monochromatic plane wave perpendicularly incident on the boundary 1–2, and calling E_i the incoming electric field, the first contribution to the reflected field is: $E_{r1} = \rho_{12} E_i$. The transmitted field $E_{t1} = \tau_{12} E_i$ is partially back-reflected by the interface 2–3, producing a second contribution to the reflected field given by: $E_{r2} = \exp(i\Delta)\rho_{23}\tau_{21}E_{t1}$. This contribution is phase-shifted with respect to the first one because the wave has traveled back and forth inside layer 2. The shift is: $\Delta = 4\pi n_2 d/\lambda$, where λ is the vacuum wavelength of the incident wave. Note that, in this case, the coefficients ρ_{12} and ρ_{23} have the same sign, so that no contribution to the phase shift between E_{r1} and E_{r2} is coming from the reflections at the interfaces.

Since the wave back-reflected from the surface 2–3 is partially reflected again when it reaches the surface 2–1, the total reflected field should be calculated by adding an infinite number of multiple reflections. Assuming that the reflectance of the interfaces 1–2 and 2–3 is small, which also means that $\tau_{21}\tau_{12} \approx 1$, the summation can be truncated by taking only the first two terms:

$$E_r = E_{r1} + E_{r2} = [\rho_{12} + \exp(i\Delta)\rho_{23}]E_i \tag{2.49}$$

It is clear from (2.49) that E_r vanishes through destructive interference if E_{r2} and E_{r1} have the same amplitude, but opposite sign. The amplitudes of the two

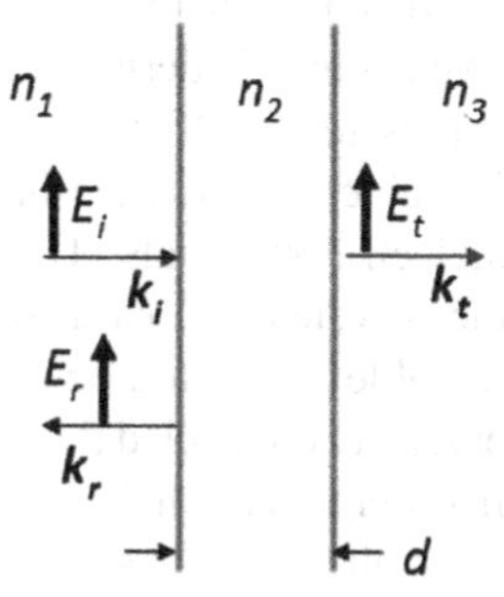

Fig. 2.4 Anti-reflection coating

fields are the same if $\rho_{12} = \rho_{23}$. According to (2.34), this is equivalent to saying that $(n_2/n_1) = (n_3/n_2)$, or: $n_2 = \sqrt{n_3 n_1}$. This means that the intermediate layer should have an index of refraction that is the geometric mean of n_1 and n_3. The two contributions have opposite sign if $\Delta = \pi$, which implies $d = \lambda/(4n_2)$. This means that the layer thickness must be a quarter of the wavelength inside the medium.

The fact that the condition imposed on the layer thickness d is wavelength-dependent indicates that there is a price to pay for eliminating reflection: the method works exactly at only one specific wavelength. In other words this anti-reflection coating is chromatic.

Instead of using the approximate expression given by (2.49), E_r can be exactly calculated by summing up an infinite number of multiple reflections or by writing the boundary conditions at the two interfaces. The latter approach is the more convenient, also because it can be more easily generalized to an arbitrary number of layers.

Consider first the interface 1–2; at left there are two waves, the incident wave with field E_i and the reflected wave with field E_r, at right there is a wave leaving the interface, with field E'_t and one arriving at the interface, with field E'_r. The boundary conditions are:

$$\begin{aligned} E_i + E_r &= E'_t + E'_r \\ E_i - E_r &= (n_2/n_1)(E'_t + E'_r). \end{aligned} \tag{2.50}$$

Taking into account that the wave crossing layer 2 acquires a phase shift of $\Delta/2$, the following boundary conditions are written for the interface 2–3:

$$\begin{aligned} \exp(i\Delta/2)E'_t + \exp(-i\Delta/2)E'_r &= E_t \\ \exp(i\Delta/2)E'_t + \exp(-i\Delta/2)E'_r &= (n_3/n_2)E_t. \end{aligned} \tag{2.51}$$

Equations (2.50) and (2.51) constitute a linear system of four equations in four variables that can be exactly solved. The expression obtained for E_r is:

$$E_r = \rho_{12}E_i + \rho_{23}E_i \frac{\tau_{12}\tau_{21}\exp(i\Delta)}{1 - \rho_{23}\rho_{21}\exp(i\Delta)} \tag{2.52}$$

Recalling that $\tau_{12}\tau_{21} = 1 - \rho_{12}^2$, it is immediately apparent that the conditions for a vanishing E_r derived from (2.52) coincide with those imposed by the approximate expression (2.49).

Table 2.2 gives the index of refraction and the transparency window of several dielectric materials. The index of refraction of the thin film depends not only on the wavelength, but also on the preparation method, and so the values of n listed in Table 2.2 should be only considered as indicative. The deposition of layers with precisely controlled thickness is usually performed by vacuum evaporation methods. In order to have an idea of the required thickness, consider the anti-reflection coating of an air-glass surface. Taking $\lambda = 550$ nm, $n_1 = 1$ (air), and $n_3 = 1.52$ (glass), the

Table 2.2 Properties of materials used for deposition of thin films

Material	n	Transparency window (μm)
Criolite Na_3AlF_6	1.35	0.15–14
Magnesium fluoride MgF_2	1.38	0.12–8
Silicon oxide SiO_2	1.46	0.17–8
Aluminum oxide Al_2O_3	1.62	0.15–6
Silicon monoxide SiO	1.9	0.5–8
Zirconium oxide ZrO_2	2.00	0.3–7
Cerium oxide CeO_2	2.2	0.4–16
Titanium oxide TiO_2	2.3	0.4–12
Zinc sulfide ZnS	2.32	0.4–14
Cadmium telluride $CdTe$	2.69	0.9–16
Silicon Si	3.5	1.1–10
Germanium Ge	4.05	1.5–16
Lead telluride $PbTe$	5.1	3.9–20

coating should be made with a material having $n_2 = \sqrt{1.52} = 1.23$ and thickness $d = 550/(4n_2) = 112$ nm. In practice, given the list of available materials in Table 2.2, the best choice would be magnesium fluoride. A quarter-wave layer of MgF_2 reduces the reflectance from 4 to 1.3 %.

By using more than one layer it is possible to cancel completely the reflectance of the air-glass surface. For instance, the sequence of two quarter-wave layers with refractive indices n_2 and n_2', produces a vanishing reflectance if $n_2'/n_2 = \sqrt{n_3/n_1}$. Putting $n_1 = 1$ e $n_3 = 1.52$, one finds $n_2'/n_2 = 1.23$. Zero-reflectance is obtained by choosing from Table 2.2 ZrO_2 ($n_2' = 2.00$) and Al_2O_3 ($n_2 = 1.62$).

Up to now normal incidence has been assumed. What happens if θ_i is different from 0, as shown in Fig. 2.5? The phase difference between the first and the second reflection becomes:

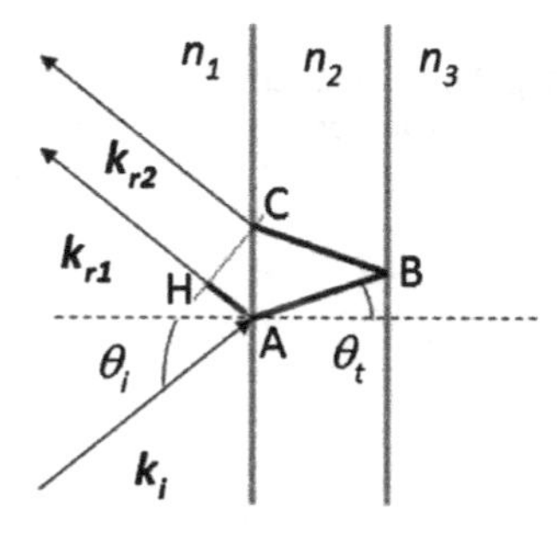

Fig. 2.5 Anti-reflection coating: non-normal incidence. The optical path difference between $\mathbf{k}_{r2}$ and $\mathbf{k}_{r1}$ is: $2n_2\overline{AB} - n_1\overline{AH}$

$$\Delta' = 2\frac{2\pi n_2}{\lambda}\frac{d}{\cos\theta_t} - \frac{2\pi n_1 d}{\lambda} tg\theta_t \sin\theta_i = \frac{4\pi n_2 d}{\lambda}\cos\theta_t, \quad (2.53)$$

where θ_t, the propagation angle inside medium 2, is related to θ_i by Snell's law. It is important to note that, once d and n_2 are fixed, the wavelength at which the reflectance vanishes, corresponding to $\Delta' = \pi$, depends on θ_i.

2.2.4 Multilayer Dielectric Mirror

By using a sequence of dielectric layers it is possible to obtain high reflectance at any desired wavelength. An important advantage of multilayer dielectric mirrors, as compared to metallic mirrors, is that there is no absorption. This means they can stand high intensities of visible light without being damaged. An important difference is that metallic mirrors have a broad-band reflectance, whereas the reflectance of dielectric mirrors is narrow-banded.

The structure of a multilayer dielectric mirror is sketched in Fig. 2.6: between medium 1 (air) and medium 2 (glass substrate) there is a sequence of alternating layers ABAB..A with refractive indices $n_A > n_B$, and $n_A > n_2$.

Clearly, the maximum reflectance will be obtained when the fields reflected by all the interfaces add up in phase at the first interface, so that constructive interference is exploited. The field reflected at the interface 1-A is phase-shifted by π with respect to the incident field because $n_A > n_1$. The field reflected at the first A–B interface does not suffer any phase-shift because $n_A > n_B$, but, upon arriving at the interface 1-A, it will acquire the phase shift $4\pi n_A d_A/\lambda$ because of the propagation back and forth within the layer thickness d_A. This second reflection will be in phase with the first one if the layer thickness corresponds to a quarter wavelength: $d_A = \lambda/(4n_A)$. By analyzing all subsequent reflections in a similar way, it is immediately clear that the maximum reflectance, $R_{\max}$, is obtained if all the layers A and B have a quarter wavelength thickness. The number $2N + 1$ of layers is always an odd number, in order to ensure that the last reflection (occurring at the A-2 interface) is also in phase with the previous ones.

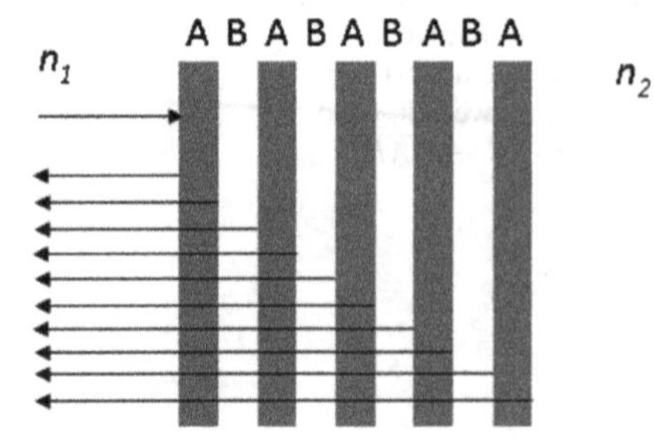

Fig. 2.6 Multilayer dielectric mirror

A detailed calculation can be developed by writing the boundary conditions at each interface, and noting that each pair of boundary conditions corresponding to a specific interface, such as (2.50) and (2.51), can be expressed in terms of a matrix equation. Therefore each interface introduces a transmission matrix, and the solution is always expressible as a cascade of matrices, independently from the number of layers.

The maximum reflectance at normal incidence is found to be:

$$R_{\max} = \left[\frac{1 - (n_A/n_1)(n_A/n_2)(n_A/n_B)^{2N}}{1 + (n_A/n_1)(n_A/n_2)(n_A/n_B)^{2N}} \right]^2. \tag{2.54}$$

Note that an expression of the type

$$\left(\frac{1-a}{1+a} \right),$$

can be transformed, by introducing the new variable $b = \ln a$, into

$$\left(\frac{1-a}{1+a} \right) = \left(\frac{e^{-b/2} - e^{b/2}}{e^{-b/2} + e^{b/2}} \right) = \tanh \left(\frac{b}{2} \right).$$

Therefore $R_{\max}$ can be expressed as:

$$R_{\max} = \tanh^2(N\xi), \tag{2.55}$$

where:

$$\xi = \ln \left(\frac{n_A}{n_B} \right) + \frac{1}{2N} \left[\ln \left(\frac{n_A}{n_1} \right) + \ln \left(\frac{n_A}{n_2} \right) \right] \tag{2.56}$$

By increasing N, $R_{\max}$ is increased and, at the same time, the high-reflectance band becomes narrower and narrower. In principle, as N tends towards infinity, $R_{\max}$ tends to 1. The greater is the ratio of n_A/n_B, the quicker $R_{\max}$ converges to its maximum value. For example, alternating zinc sulfide ($n_A = 2.32$) and magnesium fluoride ($n_B = 1.38$) layers, a reflectance of $R = 0.99$ can be reached using only 13 layers.

Similarly to the case of the anti-reflection coating, analogous calculation of the reflectance of the multilayer mirror can be performed for any incidence angle, provided the propagation path inside each layer is appropriately expressed. The dielectric multilayer mirror offers a great flexibility because it can be designed for any value of reflectance at any wavelength and any incidence angle. In particular, beam splitters having any arbitrary power partition between transmitted and reflected beam can be made.

The idea that the optical properties of a periodic structure can be considerably varied not by changing materials, but by simply acting on the geometry has gained more and more importance in the last decade. Two-dimensional and three-dimensional periodic structures made of dielectric or semiconductor materials, called photonic crystals, are increasingly utilized in photonic devices. It is interesting to note that this kind of phenomenon occurs also in Nature. For example, the brilliant colors of some butterflies are due not to absorption, but to the presence of periodic dielectric layers on their wings. A phenomenon typical of periodic structures is iridescence, due to the fact that the wavelength of maximum reflectance depends on the incidence angle. This pleasing visual effect, also called opalescence, can be observed with opals, semi-precious stones consisting of a three-dimensional periodical arrangement of sub-micrometric silica spheres.

2.2.5 Beam Splitter

As mentioned in the previous section, beam splitters are semi-reflecting mirrors that divide an optical beam into two parts propagating in different directions. In order to understand the behavior of some optical systems utilizing beam splitters, it is useful to know the phase relation between transmitted and reflected fields.

Let $\tau = \tau_o \exp(i\phi_\tau)$ and $\rho = \rho_o \exp(i\phi_\rho)$ be, respectively, the transmission and reflection coefficients for the electric field. The actual value taken by these coefficients depends on many parameters, such as the beam-splitter structure, the incidence angle, and the wavelength of the incident beam. On the other hand, as will be shown, the phase difference $\phi_\tau - \phi_\rho$ is expressed by a general law.

The scheme of Fig. 2.7 shows a beam-splitter symmetrically illuminated on both sides by two identical light beams having the same phase, polarization, incidence angle, and intensity I_o. Calling I_1 and I_2 the intensities of the two output beams and assuming no losses, it should be expected, for symmetry reason, that $I_1 = I_2 = I_o$. Since the field E_1 at output 1 is the sum of two contributions, one coming from reflection and the other from transmission, $E_1 = \rho E_o + \tau E_o$, the intensity at output 1 is given by: $I_1 = |\rho + \tau|^2 I_o$. The condition $I_1 = I_o$ is satisfied if:

$$|\rho + \tau|^2 = [\tau\tau^* + \rho\rho^* + 2Re(\tau\rho^*)] = 1. \tag{2.57}$$

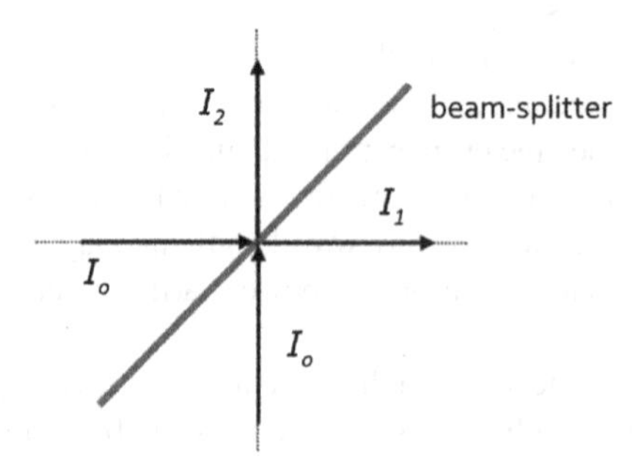

Fig. 2.7 Two identical light beams impinging on the two sides of a beam splitter

Since $\tau\tau^* + \rho\rho^* = 1$, the quantity $\mathrm{Re}(\tau\rho^*) = \tau_o\rho_o\cos(\phi_\tau - \phi_\rho)$ must be nil, therefore:

$$\phi_\tau - \phi_\rho = \frac{\pi}{2}. \tag{2.58}$$

Equation (2.58) is a general relation, valid for any lossless beam-splitter, regardless of the structure of the mirror or the incidence angle. The only limitation is that the beam-splitter should be symmetrical. It is easy to generalize to the non-symmetrical case: calling ϕ_τ and ϕ_ρ the phase shifts for a beam incident to the left of the mirror, and ϕ'_τ and ϕ'_ρ the phase shifts for a beam incident to the right of the mirror, instead of (2.58), the following relation is found:

$$\phi_\tau + \phi'_\tau - \phi_\rho - \phi'_\rho = \pi. \tag{2.59}$$

2.2.6 *Total-Internal-Reflection Prism*

The total-internal-reflection glass prism, shown in Fig. 2.8, behaves as a 100 %-reflectance mirror if the incident beam undergoes total reflections at the two glass-air interfaces. This happens if the incidence angles at the glass-air boundaries are larger than the limit angle. In the case of the figure, $\theta_i = 45°$, which is larger than θ_c for a normal optical glass. The spurious reflection of about 4 % at the air-glass input interface can be eliminated by an anti-reflection coating.

Whereas multilayer dielectric mirrors present high reflectance in a narrow wavelength band, the total-internal-reflection mirror is a broadband component. The only limitations could come from the decrease of the refractive index of the glass as the wavelength is increased, which could move θ_c to a value larger than 45°.

An interesting property of the total-internal-reflection prism is that the reflected beam is always propagating in the opposite direction with respect to the incident beam, whatever is the angle of incidence, provided that the propagation vector of the beam lies in a plane perpendicular to the prism axis (this is the plane of Fig. 2.8). Such a property is not shared by normal mirrors. Instead of a prism, it is possible to use a trihedral (see Fig. 2.9), known as the corner cube. With this, the beam is retro-reflected after three internal reflections, and it is possible to obtain a back-propagating reflected beam for any direction of incidence. In practical applications,

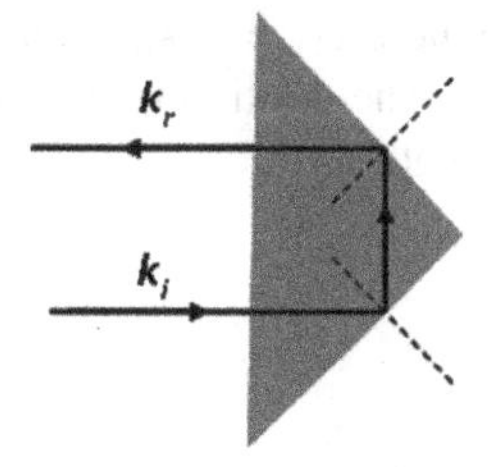

Fig. 2.8 Total internal reflection prism

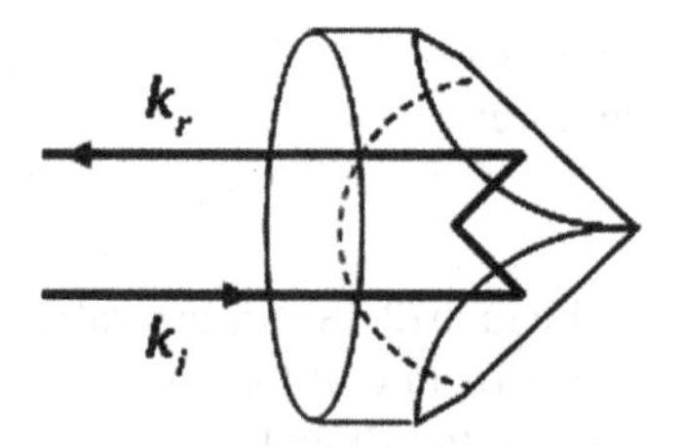

Fig. 2.9 Corner cube

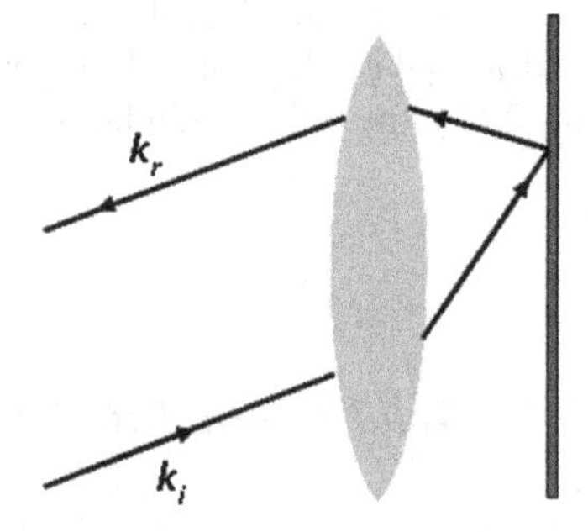

Fig. 2.10 Cat's eye

the back-reflectors placed on vehicles or on wayside posts have a structure imitating an array of corner cubes.

This same property of the corner cube is also shared by the combination of a lens plus a mirror placed in the focal plane of the lens, known as a cat's eye (see Fig. 2.10).

2.2.7 Evanescent Wave

A linearly polarized beam impinges on the glass-air planar boundary with an incidence angle $\theta_i \geq \theta_c$. Assume that the boundary coincides with the x-y plane, that the propagation vector $\mathbf{k_1}$ of the incident beam lies in the y-z plane, and that the beam is polarized along x as shown in Fig. 2.11a. Using the subscript 1 for glass and 2 for air, the electric fields of the incident and transmitted waves can be written as:

$$\begin{aligned}\mathbf{E_i} &= E_i \exp[-i(\omega t - k_{1y}y - k_{1z}z)]\mathbf{x} \\ \mathbf{E_t} &= E_t \exp[-i(\omega t - k_{2y}y - k_{2z}z)]\mathbf{x},\end{aligned} \tag{2.60}$$

where $k_{1y} = -k_o n_1 \sin\theta_i$ and $k_{1z} = k_o n_1 \cos\theta_i$.

The continuity condition at the boundary $z = 0$ imposes $k_{1y} = k_{2y}$. Using the relation:

$$k_2 = k_o n_2 = \sqrt{{k_{2y}}^2 + {k_{2z}}^2}, \tag{2.61}$$

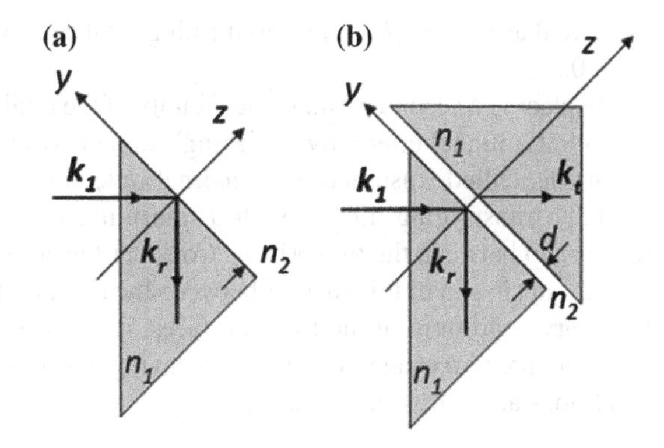

Fig. 2.11 **a** Geometry for total internal reflection. **b** Transmission of the evanescent wave by a pair of prisms

it is found: $k_{2z} = k_o\sqrt{n_2{}^2 - n_1{}^2\sin^2\theta_i}$. If $\theta_i > \theta_c$, the term under square root is negative, so that k_{2z} becomes imaginary. Putting $\sqrt{n_2{}^2 - n_1{}^2\sin^2\theta_i} = i\Gamma n_2$, where Γ is a real quantity, it is found:

$$\mathbf{E_t} = E_t\exp[-i(\omega t + k_o n_1 \sin\theta_i y)]\exp(-\Gamma k_o n_2 z)\mathbf{x}. \tag{2.62}$$

The condition of total internal reflection at the boundary between glass and air does not imply that there is no electromagnetic field in the air beyond the glass. Indeed Eq. (2.62) describes a wave, known as evanescent wave, running parallel to the glass-air interface, with an amplitude that is an exponentially decreasing function of the penetration distance z. The electromagnetic energy density inside medium 2, being proportional to $|E_t|^2$, is also exponentially decreasing with z, becoming smaller by a factor e^{-1} at the distance:

$$\delta = \frac{1}{2\Gamma k_o n_2} = \frac{\lambda}{4\pi\sqrt{n_1{}^2\sin^2\theta_i - n_2{}^2}}. \tag{2.63}$$

For instance, if $n_1 = 1.5$, $n_2 = 1$, $\theta_i = \pi/4$, it follows that $\Gamma = 0.354$, therefore $\delta = \lambda/4.44$. Taking the central wavelength of the visible range, $\lambda = 0.55\ \mu\text{m}$, it is found that $\delta = 120$ nm.

In order to derive the amplitude of the transmitted and reflected fields, the following boundary conditions are written:

$$E_i + E_r = E_t \tag{2.64}$$

$$E_i - E_r = i\gamma E_t, \tag{2.65}$$

where $\gamma = \Gamma n_2/(n_1\cos\theta_i)$ is a real quantity. The solutions of (2.64) and (2.65) are:

$$E_t = \frac{2}{1+i\gamma}E_i\ ; \quad E_r = \frac{1-i\gamma}{1+i\gamma}E_i. \tag{2.66}$$

Note that $|E_i| = |E_r|$, thus confirming total reflectance, even if $|E_t|$ is different from 0.

By placing a second prism in the vicinity of the totally reflecting one, it is possible to exploit a tunnel effect, by "catching" some part of the evanescent wave. Such a technique, called frustrated total internal reflection, allows for the realization of a partially transmitting mirror, with a transmittance controlled by the thickness of the air-gap between the two prisms. Consider the configuration in Fig. 2.11b, where there is an air-gap of thickness d between the two prisms. In order to write down the boundary conditions at the two interfaces, it is necessary to assume that inside the gap there are two evanescent waves decaying in opposite directions. The continuity conditions at the first interface are:

$$E_i + E_r = E_t' + E_r', \tag{2.67}$$

and

$$E_i - E_r = i\gamma(E_t' - E_r'), \tag{2.68}$$

where E_t' and E_r' are the fields of the two evanescent waves, the first decaying as a function of z and the second growing as a function of z. Similarly, the continuity conditions at the second interface are:

$$\exp[-d/(2\delta)]E_t' + \exp[d/(2\delta)]E_r' = E_t, \tag{2.69}$$

and

$$\exp[-d/(2\delta)]E_t' - \exp[d/(2\delta)]E_r' = -(i/\gamma)E_t. \tag{2.70}$$

By eliminating E_t' and E_r', E_t and E_r can be calculated. In particular, the transmittance T is found to be:

$$T = \frac{|E_t|^2}{|E_i|^2} = \frac{\exp(-d/\delta)}{\left[\Big(1 + \exp(-d/\delta)\Big)/2\right]^2 + A\Big(1 - \exp(-d/\delta)\Big)^2}, \tag{2.71}$$

where $A = (1-\gamma^2)^2/(16\gamma^2)$. It is seen from (2.71) that, for d tending to 0, T tends to 1, as expected. If $d \gg \delta$, T is approximated by:

$$T \approx \frac{4\exp(-d/\delta)}{1+4A}, \tag{2.72}$$

which shows an exponential decay of T as a function of d. The set of two prisms is therefore equivalent to a variable-transmittance attenuator.

2.2.8 Thin Lens and Spherical Mirror

So far in this chapter only situations involving plane waves and plane boundaries have been considered. In this section spherical boundaries are discussed, concentrating the attention on the important optical component of the lens.

The lens, as shown in Fig. 2.12, is formed by joining two spherical caps made of glass. The radii of curvature of the two surfaces, ρ_1 and ρ_2, are conventionally taken to be positive if the surfaces are convex. The thin-lens approximation consists in assuming that the lens thickness is much smaller than the radii of curvature of the two spherical surfaces. Neglecting reflections at the two surfaces, the effect of the lens on the incident field E_1 is described by a transmission function $\tau_f(x, y)$:

$$E_2(x, y, z, t) = \tau_f(x, y)E_1(x, y, z, t). \tag{2.73}$$

This transmission function has a unitary modulus and a phase that is locally dependent on the glass thickness:

$$\tau_f(x, y) = \exp\{ik_o[n\Delta(x, y) + \Delta_o - \Delta(x, y)]\}, \tag{2.74}$$

where n is the glass refractive index, $\Delta(x, y)$ is the lens thickness at the transversal position (x, y), and $\Delta_o = \Delta(0, 0)$ is the thickness on the optical axis.

The phase shift is the sum of two contributions, one due to the path inside the glass, $k_o n\Delta(x, y)$, and the other due to the path in air, $k_o[\Delta_o - \Delta(x, y)]$. The lens thickness at the generic point (x, y) is:

$$\Delta(x, y) = \Delta_o + \sqrt{\rho_1{}^2 - (x^2 + y^2)} - \rho_1 + \sqrt{\rho_2{}^2 - (x^2 + y^2)} - \rho_2 \tag{2.75}$$

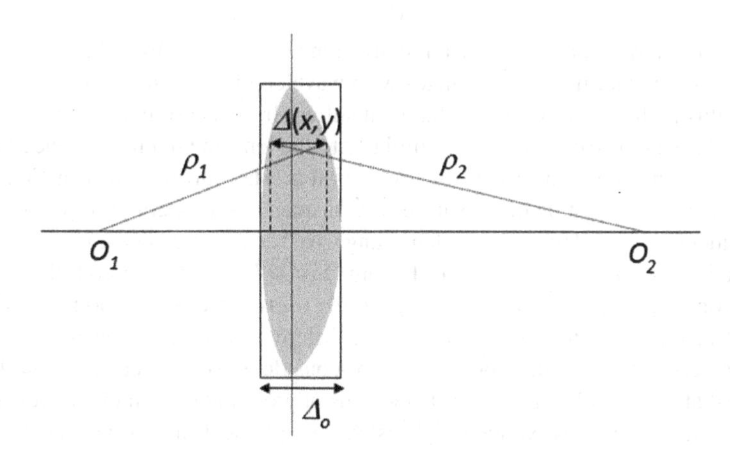

Fig. 2.12 Thin lens made of two glass caps with radii of curvature ρ_1 and ρ_2

Equation (2.75) can be simplified by assuming that the lens diameter is much smaller than the radii of curvature of the two surfaces, so that:

$$\sqrt{\rho_1{}^2 - (x^2 + y^2)} = \rho_1 \sqrt{1 - \frac{x^2 + y^2}{\rho_1{}^2}} \approx \rho_1 \left[1 - \frac{x^2 + y^2}{2\rho_1{}^2} \right], \tag{2.76}$$

and

$$\sqrt{\rho_2{}^2 - (x^2 + y^2)} \approx \rho_2 \left[1 - \frac{x^2 + y^2}{2\rho_2{}^2} \right]. \tag{2.77}$$

In the power expansion of the square-root terms, only the first-order terms in $(x^2 + y^2)$ are kept. This is equivalent to approximate the spherical surface with a paraboloidal surface, similarly to the paraxial approach used in Chap. 1 for the description of spherical waves.

Inserting (2.76) and (2.77) into (2.75):

$$\Delta(x, y) = \Delta_o - \frac{x^2 + y^2}{2} \left(\frac{1}{\rho_1} + \frac{1}{\rho_2} \right). \tag{2.78}$$

By substituting inside (2.74):

$$\tau_f(x, y) = \exp(ik_o n \Delta_o) \exp \left[-ik_o \frac{x^2 + y^2}{2f} \right], \tag{2.79}$$

where f, called focal length, is defined by the relation:

$$\frac{1}{f} = (n - 1) \left(\frac{1}{\rho_1} + \frac{1}{\rho_2} \right) \tag{2.80}$$

Note that f has a sign: $f > 0$ means converging lens, $f < 0$ diverging lens. Since f depends on n, and, in turn, n changes with wavelength, the lens focuses different wavelengths to different distances, that is, it suffers from a chromatic aberration.

The spherical mirror is described similarly to the lens, the main difference being that it works in reflection instead of transmission. The mirror shown in Fig. 2.13 has center in O and radius of curvature ρ. Consider a beam parallel to the optical axis, incident at point P with an incidence angle α. The reflected beam intersects the optical axis at point F. The distance of F from O is: $\overline{OF} = \overline{OP}/(2\cos\alpha)$. Since $\overline{OF}$ depends on α, parallel beams impinging on the mirror in different points produce reflected beams crossing the optical axis at different positions. However, if only beams traveling close to the optical axis are considered (i.e. where $\cos\alpha \approx 1$), it is found that $\overline{OF} \approx \overline{OP}/2 = \rho/2$. Under this approximation, which is equivalent to treating the surface as paraboloidal instead of spherical, it can be said that the spherical mirror brings all parallel beams to converge at a single point a distance $\rho/2$ from the mirror surface. In other words, the mirror behaves in reflection just as a lens

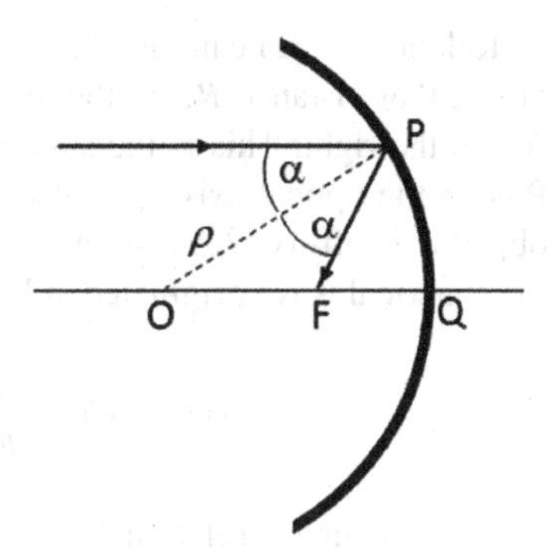

Fig. 2.13 Spherical mirror with radius of curvature ρ

having a focal length $f = \rho/2$. Note that the focal length of the spherical mirror is independent of the wavelength. All large astronomical telescopes use parabolic mirrors instead of lenses.

2.2.9 Focusing Spherical Waves

A spherical wave traveling along the z axis impinges on a thin lens located at the plane $z = 0$. If R_1 is the radius of curvature of the wavefront, then, under the paraxial approximation, the electric field E_1 of the wave at $z = 0$ is given by:

$$E_1(x, y, 0) = \frac{A_o}{R_1}\exp\left(ik_o\frac{x^2 + y^2}{2R_1}\right). \tag{2.81}$$

The field distribution transmitted through the lens, $E_2 = \tau_f E_1$, calculated by using (2.79) and (2.81), is again that of a spherical wave, with the modified radius of curvature R_2, which is related to R_1 by:

$$\frac{1}{R_2} = \frac{1}{R_1} - \frac{1}{f}. \tag{2.82}$$

If $R_1 = f$, R_2 becomes infinite, which means that a spherical wave originating from the left focus is converted into a parallel beam (plane wave). If $R_1 > f$, $R_2 < 0$, indicating that the diverging wave is converted into a converging wave. If $R_1 \gg f$, $R_2 \approx f$, that is, a plane wave is converted into a spherical wave converging at the right focal plane. In practice, the wave will not converge to a point, but the beam spot at focus will be described by an Airy function (see Sect. 1.6.1) because of the diffraction effect introduced by the finite dimension of the lens-diameter.

The ratio between the lens radius and the focal length is called numerical aperture (NA) of the lens. It describes the light-gathering capacity of the lens.

Relation (2.82) can also be interpreted in the following way: given a point-like object P at distance R_1 on the left, the lens creates a real image Q of P at distance R_2 on the right. This is the standard interpretation given by geometrical optics. If P is on the optical axis, Q is also on the same axis. What happens if the point-like object P′ is off-axis? Assigning to P′ the coordinates $(x_o, 0, R_1)$, the electric field of the spherical wave originated at P′, evaluated at the lens plane $z = 0$, is now:

$$E_1(x, y, 0) = \frac{A_o}{R_1}\exp\left\{\frac{ik_o}{2R_1}[(x - x_o)^2 + y^2]\right\}. \quad (2.83)$$

Using again the relation $E_2 = \tau_f E_1$, after some algebra the following expression is found for E_2:

$$E_2(x, y, 0) = \frac{A_o\exp(i\phi)}{R_1}\exp\left\{\frac{ik_o}{2R_2}\left[(x + \frac{f}{R_1 - f}x_o)^2 + y^2\right]\right\}, \quad (2.84)$$

where ϕ depends on x_o and on the lens parameters, but not on x and y. Equation (2.84) shows that the image of P′ is at point Q′ having coordinates $[-f x_o/(R_1 - f), 0, R_2]$. It can be said that the lens creates an image QQ′ of the segment PP′. If $R_1 > f$, the real image is upside-down, with a magnification $f/(R_1 - f)$. This approach is therefore equivalent to the geometrical-optics treatment. The topic of image formation has important applications but will not be further discussed in this text.

Another case to be considered is that of a plane wave having a propagation direction slightly tilted with respect to the optical axis. To simplify the discussion it is assumed that the wave vector $\mathbf{k}$ is in the plane x-z, forming an angle α with x. At $z = 0$ the incident field is expressed as $E_1 = A_o\exp(ik_o\alpha x)$, where $\sin\alpha$ has been approximated as α. After using $E_2 = \tau_f E_1$ again, and applying some algebra, it is found that:

$$E_2(x, y, 0) = A_o\exp(i\phi)\exp\left\{-\frac{ik_o}{2f}\left[(x - \alpha f)^2 + y^2\right]\right\} \quad (2.85)$$

where $\phi = k\alpha^2 f/2$. Equation (2.85) describes a spherical wave that converges at the point $(\alpha f, 0, f)$ located on the right focal plane, as shown in Fig. 2.14.

Generalizing the treatment, it can be stated that a biunique correspondence is established between the propagation direction of the incident wave and points on the right focal plane.

2.2.10 Focusing Gaussian Spherical Waves

If the wave impinging on the lens is a Gaussian spherical wave with a complex radius of curvature of the wavefront q_1, then it can be immediately shown, by using (2.73) and (2.79), that the transmitted wave is still a Gaussian spherical wave. The new

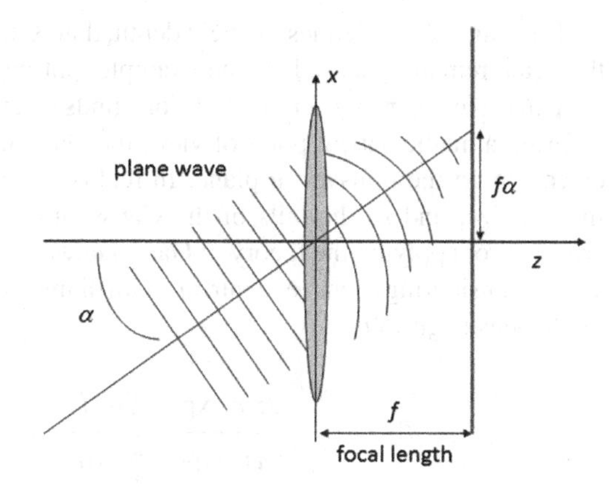

Fig. 2.14 A plane wave corresponds to a point at the focal plane of a lens

complex radius of curvature q_2 satisfies the relation:

$$\frac{1}{q_2} = \frac{1}{q_1} - \frac{1}{f}. \tag{2.86}$$

Since the beam radius is unchanged by crossing the lens, $w_1 = w_2$, one finds that (2.86) coincides with (2.82). As an example, if $f > 0$ and E_1 describes a diverging Gaussian wave with $R_1 > f$, the transmitted Gaussian wave has a negative radius of curvature of the wavefront, $R_2 < 0$, and is, therefore, converging. There is a basic difference between this case and the one treated in the previous section: the theory of Gaussian waves accounts intrinsically for diffraction effects, such that the wave converges to a finite-size beam spot.

Once q_2 at $z = 0$ is assigned, it is possible to obtain the distance z_f at which the focused beam takes the minimum radius, and also the minimum radius w_f, by using relations 1.35 and 1.36. The results of the calculation are:

$$z_f = \frac{|R_2|}{1 + [\lambda |R_2|/(\pi w_1^2)]^2}, \quad w_f = \frac{w_1}{\sqrt{1 + [\pi w_1^2/(\lambda |R_2|)]^2}}. \tag{2.87}$$

Assuming that $R_1 \gg f$, which means $|R_2| \approx f$, and $\pi w_1^2 \gg \lambda f$, the relations of (2.87) simplify as follows:

$$z_f \approx f, \;\; w_f \approx \frac{\lambda f}{\pi w_1} \tag{2.88}$$

It is useful to calculate the Rayleigh length at focus:

$$z_{fR} = \frac{\pi w_f^2}{\lambda} \approx \frac{\lambda}{\pi} \left(\frac{f}{w_1} \right)^2. \tag{2.89}$$

The length $2z_{fR}$ defines the field depth, that is, the interval of distance over which the beam remains focused. As an example, putting $f = 20$ cm, $w_1 = 2$ mm, $\lambda = 0.5\,\mu$m (note that $\pi w_1^2/(\lambda f)$ is $\gg 1$), one finds that $w_f = 16\,\mu$m and $z_{fR} = 1.6$ mm.

From a mathematical point of view, the Gaussian spherical wave has an infinite extension on the transversal plane. In reality, practical optical components have a limited size, and so the tails of the Gaussian distribution are lost. The empirical criterion for applying the theory of Gaussian waves is if the lost power is sufficiently small. Considering a lens (or a mirror) with diameter d, the power fraction intercepted by the lens is given by:

$$T_f = \frac{\int_0^{d/2} 2\pi r \exp(-\frac{2r^2}{w^2})dr}{\int_0^{\infty} 2\pi r \exp(-\frac{2r^2}{w^2})dr} = 1 - \exp\left(-\frac{d^2}{2w^2}\right) \tag{2.90}$$

As an example, if $d = 2w$, the intercepted power fraction is 86 %, if $d = 3w$, $T_f = 0.99$. Therefore, if $d > 3w$, it is reasonable to assume the applicability of the Gaussian wave theory.

2.2.11 Matrix Optics

Within the geometrical optics approximation, a useful approach is that of associating a 2×2 matrix, called $ABCD$ matrix, with each optical component.

The ray present at the coordinate z is characterized by two parameters, the distance $r(z)$ from the optical axis, and the angle its direction forms with the optical axis, $\phi(z)$. Within the paraxial approximation, $\phi(z)$ is simply the derivative of r with respect to z. The ray is represented by a one-column matrix:

$$\mathbf{R} = \begin{pmatrix} r \\ \phi \end{pmatrix}. \tag{2.91}$$

Every optical component is represented by a 2×2 matrix of the type:

$$\mathbf{M} = \begin{pmatrix} A & B \\ C & D \end{pmatrix}, \tag{2.92}$$

connecting the output vector $\mathbf{R_u}$ to the input vector $\mathbf{R_i}$ according to:

$$\mathbf{R_u} = \mathbf{M}\mathbf{R_i}. \tag{2.93}$$

Using refraction laws and the properties of lenses and mirrors discussed in the preceding sections, the $ABCD$ matrices of the components shown in Fig. 2.15 can be written as follows.

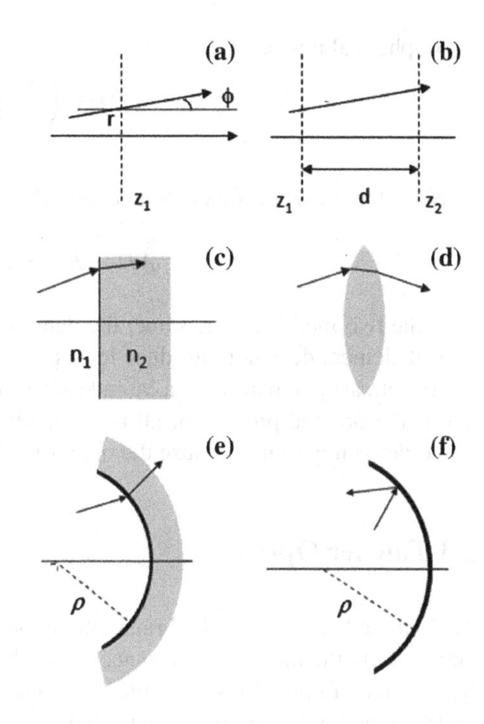

Fig. 2.15 Different optical components that are described by ABCD matrices: **a** generic ray, **b** free propagation, **c** planar boundary between two different media, **d** lens, **e** spherical boundary between two media, **f** spherical mirror

- Free propagation

$$\mathbf{M} = \begin{pmatrix} 1 & d \\ 0 & 1 \end{pmatrix} \tag{2.94}$$

- Lens

$$\mathbf{M} = \begin{pmatrix} 1 & 0 \\ \frac{-1}{f} & 1 \end{pmatrix} \tag{2.95}$$

- Planar boundary between two media

$$\mathbf{M} = \begin{pmatrix} 1 & 0 \\ 0 & \frac{n_1}{n_2} \end{pmatrix} \tag{2.96}$$

- Spherical boundary between two media

$$\mathbf{M} = \begin{pmatrix} 1 & 0 \\ \frac{n_1 - n_2}{n_2} \frac{1}{\rho} & \frac{n_1}{n_2} \end{pmatrix} \tag{2.97}$$

- Spherical mirror

$$\mathbf{M} = \begin{pmatrix} 1 & 0 \\ \frac{-2}{\rho} & 1 \end{pmatrix} \tag{2.98}$$

Note that the determinant of a generic $ABCD$ matrix is given by:

$$AD - BC = n_1/n_2. \tag{2.99}$$

If the two media are the same, the determinant is equal to 1, as expected if the optical element does not introduce losses.

An optical system made by a cascade of various elements is represented by a matrix that is the ordered product of all the individual matrices describing each element. The order is important because the matrix product is generally non-commutative.

2.3 Fourier Optics

As discussed in Sect. 1.6, the Fraunhofer approximation of the diffraction integral is only valid in the far field, at distances from the screen that are much larger than the typical size of optical instruments. Therefore, it is of great practical importance to find a method of transporting the far-field behavior to short distances. In this section it will be shown that the insertion of a lens can solve the problem.

Consider a thin lens having a focal distance f and positioned at $z = 0$, that is illuminated from the left by a wave propagating along the z axis. By assuming that the field distribution in the left focal plane of the lens is known, $E_i(x_o, y_o, -f)$, the aim of the calculation is to derive the field distribution at the right focal plane, $E_f(x_f, y_f, f)$. The calculation is performed according to the following scheme: (a) the wave is propagated from the plane $z = -f$ to the plane $z = 0$ by using Fresnel diffraction, as expressed by (1.50); (b) at $z = 0$ the field distribution is multiplied by the lens transmission function $\tau_f(x, y)$; (c) the obtained field distribution is propagated to the plane $z = f$ by using again Fresnel diffraction.

Step (a) of the calculation gives:

$$E(x, y, 0) = \frac{i \exp(ikf)}{\lambda f} \int_{-\infty}^{\infty} dx_o \int_{-\infty}^{\infty} E_i(x_o, y_o, -f) \exp\left\{\frac{ik}{2f}[(x - x_o)^2 + (y - y_o)^2]\right\} dy_o. \tag{2.100}$$

Steps (b) and (c) require calculating the integral:

$$E_f(x_f, y_f, f) = \frac{i \exp(ikf)}{\lambda f} \int_{-\infty}^{\infty} dx \int_{-\infty}^{\infty} E(x, y, 0)\tau_f(x, y) \exp\left\{\frac{ik}{2f}[(x - x_f)^2 + (y - y_f)^2]\right\} dy. \tag{2.101}$$

By inserting (2.100), and the expression of τ_f given by (2.79), into (2.101), one obtains:

$$E_f(x_f, y_f, f) = -\frac{\exp[2ikf + ik(n-1)\Delta_o]}{\lambda^2 f^2} \int_{-\infty}^{\infty} dx_o \int_{-\infty}^{\infty} dy_o E_i(x_o, y_o, -f) \int_{-\infty}^{\infty} dx$$
$$\int_{-\infty}^{\infty} dy \exp\left\{\frac{ik}{2f}[(x-x_f)^2 + (y-y_f)^2 + (x-x_o)^2 + (y-y_o)^2 - x^2 - y^2]\right\}. \quad (2.102)$$

The $dxdy$ integral is solved by writing the integrand function as the exponential of a square term. Noting that: $(x-x_f)^2+(y-y_f)^2+(x-x_o)^2+(y-y_o)^2-x^2-y^2 = (x-x_o-x_f)^2+(y-y_o-y_f)^2-2(x_o x_f + y_o y_f)$, it is finally found:

$$E_f(x_f, y_f, f) = \frac{\exp[2ikf + ik(n-1)\Delta_o]}{\lambda f}$$
$$\int_{-\infty}^{\infty} dx_o \int_{-\infty}^{\infty} dy_o E_i(x_o, y_o, -f) \exp\left[\frac{ik}{f}(x_o x_f + y_o y_f)\right]. \quad (2.103)$$

Equation (2.103) represents a very important result: it states that the field distribution at the right-hand focal plane of the lens is the two-dimensional Fourier transform of the distribution at the left-hand plane. Such a result is consistent with the consideration, developed at the end of Sect. 2.2.9, that a biunique correspondence exists between the propagation direction of the incident wave and points on the right focal plane. In fact to perform the spatial Fourier transform of the field distribution is the same as describing the field by a superposition of plane waves.

Equation (2.103) has a fundamental role in many optical processing techniques. For instance, it suggests that spatial filtering of optical signals can be performed by placing an appropriate filter at the right-hand focal plane, and reconstructing the filtered signal with a second confocal lens. An example is shown in Fig. 2.16, where a pinhole is used as a low-pass filter, i.e. a filter transmitting only low spatial frequencies.

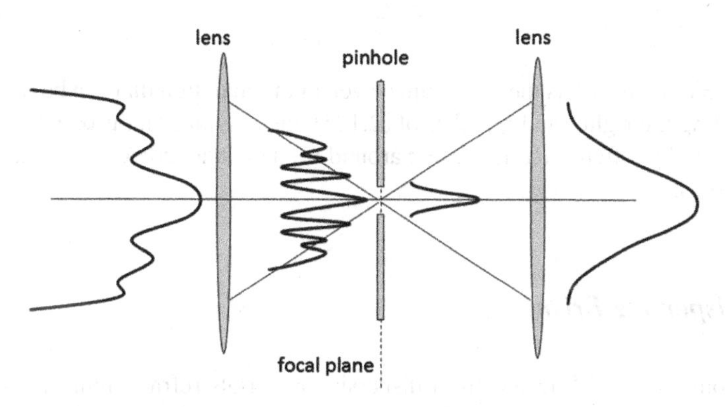

Fig. 2.16 Scheme of spatial filtering of an optical beam

2.4 Spectral Analysis

The electric field $E(t)$ associated with a generic light beam can always be described, as discussed in Sect. 2.1, as a weighted superposition of sinusoidal functions oscillating at different frequencies. The Fourier transform of $E(t)$ is the complex quantity $E(\omega)$ (given by (2.10)) that represents the amplitude and the phase of the field component at angular frequency ω.

The power spectrum of the light beam, $S(\omega)$, is defined as:

$$S(\omega) = c\frac{\varepsilon_o |E(\omega)|^2}{2} \tag{2.104}$$

It should be noted that $S(\omega)$ does not contain any information on the phase of $E(\omega)$. This means that the knowledge of $S(\omega)$ is not sufficient, in general, for reconstructing the temporal behavior of $E(t)$.

The usual method to measure the power spectrum of an optical signal is that of collimating a sample beam and sending it through a dispersive system in order to angularly separate the different frequency components. This method was introduced by Newton, who first observed the spectrum of solar light through a glass prism. Instead of using the dispersion properties of glass, angular separation can be generated by diffraction gratings, i.e. structures with periodic transmission functions. Almost all modern commercial spectrometers are based on gratings. As will be seen, there is another method, consisting of the use of a narrow-band tunable filter. This method is conceptually similar to that used to measure the power spectrum of electrical signals. In the optical case, the narrow-band filter is a Fabry-Perot interferometer.

An important parameter for assessing the performance of a spectrometer is the frequency resolution $\delta\nu$, defined as the minimum frequency separation that can be resolved by the spectrometer. Calling ν_o the central frequency of the signal to be analyzed, the resolving power of the instrument at that frequency is calculated as:

$$P_r = \frac{\nu_o}{\delta\nu} = \frac{\lambda_o}{\delta\lambda}, \tag{2.105}$$

where $\lambda_o = c/\nu_o$ and $\delta\lambda$ is the minimum wavelength separation that can be resolved. When writing the right-hand member of (2.105), the assumption $\delta\nu \ll \nu_o$ has been made. In fact, by differentiating $\nu\lambda = c$ around ν_o, it is obtained: $\lambda_o d\nu - \nu_o d\lambda = 0$, hence (2.105).

2.4.1 Dispersive Prism

Optical components fabricated from dispersive materials refract light of different wavelengths by different angles. Consider a collimated light beam that impinges on a glass prism at an incidence angle θ_i, as shown in Fig. 2.17. Since the angle

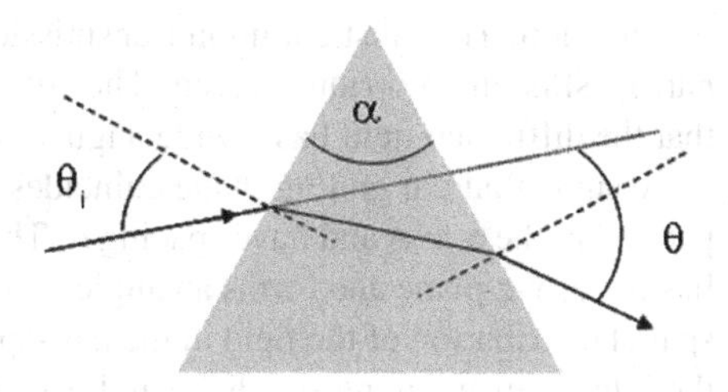

Fig. 2.17 Dispersive prism

of refraction depends on the refractive index, which is wavelength dependent, the direction of the emergent beam is also wavelength dependent.

The deflection angle θ, calculated from the refraction law, is given by:

$$\theta = \theta_i - \alpha + \arcsin\left[\sin\alpha\sqrt{n^2 - \sin^2\theta_i} - \sin\theta_i\cos\alpha\right], \tag{2.106}$$

where α is the prism angle and n the refractive index of the glass.

What's the minimum wavelength separation $\delta\lambda$ that can be resolved? Clearly, this depends on the minimum angular resolution. Even assuming that the incident beam is monochromatic and perfectly collimated, the output beam will present an angular spread due to diffraction. If L is the minimum size between prism size and incident beam diameter, and λ_o is the wavelength of the incident beam, the angular spread will be $\delta\theta \approx \lambda_o/L$. In order to resolve two different wavelengths, the difference in their deflection angles should be larger than $\delta\theta$. Therefore, the wavelength resolution is given by:

$$\delta\lambda = \left(\frac{d\theta}{d\lambda}\right)^{-1} \delta\theta, \tag{2.107}$$

where $d\theta/d\lambda$ is calculated by assigning the dependence of n on λ. Taking typical values for the parameters appearing inside (2.106), $\alpha = \theta_i = \pi/4$, $n = 1.5$, it is found that $d\theta/d\lambda \approx dn/d\lambda$. Therefore, the resolving power of the dispersive prism can be written as:

$$P_r = \frac{\lambda_o}{\delta\lambda} \approx L\frac{dn}{d\lambda} \tag{2.108}$$

If $dn/d\lambda = (10\ \mu\text{m})^{-1}$, and $L = 1$ cm, then $P_r = 10^3$.

2.4.2 Transmission Grating

The diffraction grating is an optical component that periodically modulates (in the transversal direction) the amplitude or phase of an incident wave. The treatment here is limited to linear gratings in which the diffracting elements consist of parallel lines ruled on glass or metal. These gratings may be used either in transmission or in reflection. The most common transmission grating is made of a periodic array of narrow slits on an opaque screen. The slit width is smaller than the wavelength, so that the diffracted field has a wide angular spread.

Assume that the grating plane coincides with the x-y plane, and that the slits are parallel to the x axis and have spacing d. The wave vector of the incident plane wave lies in the y-z plane and forms an angle θ_i with the z axis, as shown in Fig. 2.18. The spatial distribution of the field in the half-space beyond the grating is determined by the interference among the diffracted waves coming from all the slits. Note that in the approximation of a far-field projection the interfering beams can be considered as parallel. Diffractive interference will occur only in those directions satisfying the condition:

$$q\lambda = d(\sin\theta_q - \sin\theta_i), \tag{2.109}$$

where $q = 0, \pm 1, \pm 2, \ldots$. Equation (2.109) defines a discrete set of directions in which a diffracted beam can be observed.

In particular, the beam corresponding to the order of $q = 0$ is simply the transmitted beam: $\sin\theta_0 = \sin\theta_i$.

Equation (2.109) says that the directions of the diffraction orders (except for the order 0) are wavelength-dependent. Therefore the grating is a dispersive structure. As such it can be used, like the dispersive prism, to analyze the power spectrum of an optical signal.

Similarly to the case of the prism, the limit to the wavelength resolution comes from the diffraction effect due to the finite size of the grating. At $\theta_i = 0$ the first-order diffracted field, as derived from the theory of Fraunhofer diffraction, is expressed by (1.69). This equation shows that, in the far-field plane at distance z from the grating plane, the peak corresponding to the diffracted field is transversally displaced by

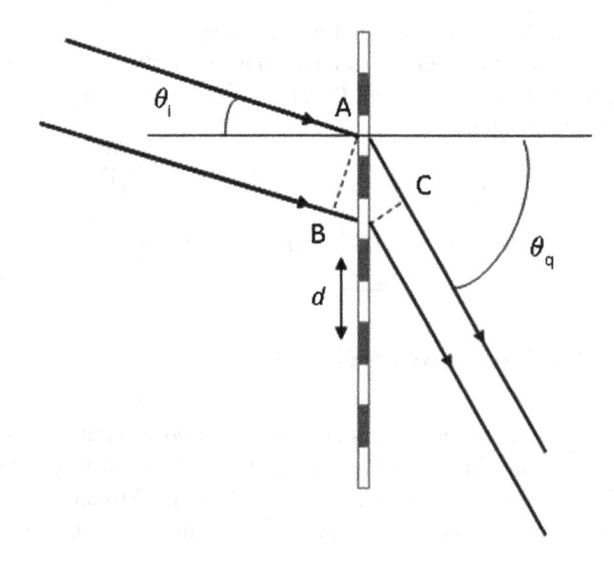

Fig. 2.18 Transmission grating. The path difference between the two diffracted beams is: $\overline{AC} - \overline{BH} = d\sin\theta_q - d\sin\theta_i$

$\lambda z/d$ and has a width $\lambda z/L$. This means that the resolving power of the grating is:

$$P_r = \frac{\lambda z/d}{\lambda z/L} = \frac{L}{d} = N, \tag{2.110}$$

where L is the size of the grating, and N the number of illuminated slits. Note that N coincides with L/d only if the collimated beam fully covers the grating surface. By introducing the number of slits per unit length, $f_o = 1/d$, $N = f_o L$. Taking, e.g., $f_o = 10^3$ mm^{-1} and $L = 1$ cm, the resolving power is 10^4, one order of magnitude larger than that of the dispersive prism. As shown by (2.109), the angular separation between two wavelengths grows by increasing the diffraction order: for the order q, the resolving power becomes $P_r = qN$. However, the fraction of diffracted optical power in a given order decreases with the value of q.

A problem arising from the use of a grating in spectral analysis is an ambiguity that can appear if the optical signal has a very broad wavelength spectrum. Indeed, considering (2.109), it can be seen that the first-order diffraction of the wavelength λ_1 appears in the same direction as the second-order diffraction of the wavelength $\lambda_1/2$. This means that there is an upper limit to the spectral width that can be unambiguously analyzed by the grating. For very broad spectra it is more convenient to use a dispersing prism.

Instead of an amplitude grating, a phase grating can be used: this is discussed in Sect. 1.6.2. Most of the properties of the amplitude grating are also shared by the phase grating.

The description of the diffraction properties of gratings assumes that the diffracted field is observed in the far field. Since, in order to have a good resolving power, the grating size must be of the order of centimeters, far field means a distance of kilometers, as discussed in Sect. 1.6. However, by exploiting the properties of lenses (or spherical mirrors) discussed in Sect. 2.3, the far field can be brought to the focal plane of a lens placed beyond the grating: this is the approach adopted by all practical spectrometers.

2.4.3 Reflection Grating

Reflection gratings are made by creating a linear alternation of reflecting and opaque zones on a surface, as shown in Fig. 2.19.

In the case of the reflection grating, the diffractive effects are observed in the backward direction. The directions of constructive interference are given by:

$$q\lambda = d(\sin\theta_i + \sin\theta_q), \tag{2.111}$$

where q is $0, \pm 1, \pm 2, \ldots$ and θ_i is the incidence angle. Inside (2.111) diffraction angles are positive if the corresponding directions stay in the upper half-plane with

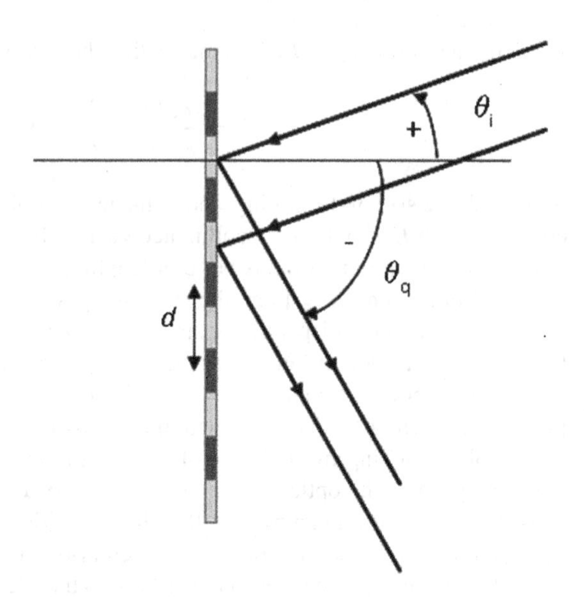

Fig. 2.19 Reflection grating. The *black* zones are opaque, and the *grey* zones are reflecting

respect to the grating plane normal. The zeroth-order beam propagates at the angle $\theta_0 = -\theta_i$, corresponding to specular reflection, that is independent from λ.

In a transmission grating the diffraction lobe due to a single slit presents maximum intensity in correspondence with the direction of incidence. As a consequence, the largest fraction of diffracted power goes to order 0. This is useless from the point of view of spectral analysis because the zeroth order is not dispersive. Similarly, in the case of a reflection grating the largest fraction of diffracted power would appear in specular reflection. Diffraction gratings can be optimized such that most of the power goes into a certain diffraction order, $q_0 \neq 0$, leading to a high diffraction efficiency for that order. This is indeed possible by using a so-called blazed grating (see Fig. 2.20), in which the reflecting elements are tilted by an angle θ_g with respect to the grating plane, obtaining a sawtooth-like profile.

The specular reflection from the tilted elements appears in a direction forming an angle $\theta_r = -(\theta_i - \theta_g) + \theta_g = (2\theta_g - \theta_i)$ with the normal to the grating plane. It is intuitive to expect that the maximum of the diffraction lobe associated to a single reflecting element occurs in the direction of specular reflection relative to that segment. By imposing that such direction coincides with that of the diffraction order q_0, $\theta_{q_o} = \theta_r$, the following relation is found:

$$\theta_g = \frac{\theta_i + \theta_{q_o}}{2}. \tag{2.112}$$

By substituting (2.112) inside (2.111):

$$q_o\lambda = d[\sin\theta_i + \sin(2\theta_g - \theta_i)]. \tag{2.113}$$

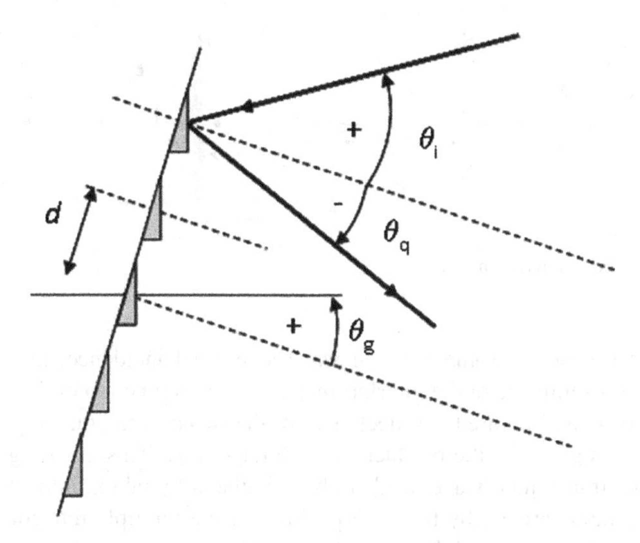

Fig. 2.20 Blazed reflection grating

Once λ and θ_i are fixed, (2.113) prescribes which is the value of θ_g to be chosen in order to maximize the power diffracted at order q_o.

There are two particularly interesting cases. The first corresponds to the Littrow configuration, in which the incident beam arrives normally to the tilted elements, $\theta_i = \theta_g$, so that (2.113) becomes:

$$q_o\lambda = 2d \sin\theta_g. \tag{2.114}$$

In this case $\theta_{q_o} = \theta_i$, which means that the order q_o propagates in the opposite direction with respect to the incident beam. The grating behaves as a wavelength-selective mirror, because it reflects only the wavelength that satisfies (2.114). As an example, consider a grating with 1200 lines per millimeter used in the Littrow configuration at the first order ($q_o = 1$) for $\lambda = 600$ nm. Equation (2.114) gives $\theta_g \approx 21°$. The back-reflected wavelength can be tuned by slightly rotating the grating, in order to change the incidence angle.

The second interesting case is that of normal incidence. If $\theta_i = 0$, $\theta_g = -\theta_{q_o}/2$. The condition of constructive interference is: $q_o\lambda = d\sin(2\theta_g)$. This is the configuration utilized in standard spectrometers.

2.4.4 Fabry-Perot Interferometer

The Fabry-Perot interferometer is a frequency filter, made by two parallel mirrors, as shown in Fig. 2.21.

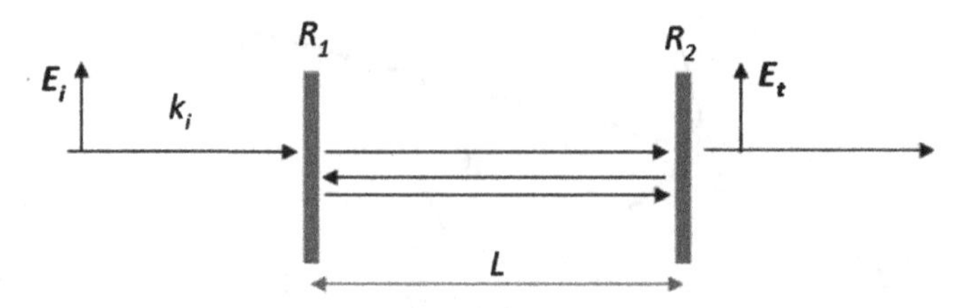

Fig. 2.21 Fabry-Perot interferometer

Considering a monochromatic plane wave at normal incidence, the aim is to calculate the transmittance of the interferometer, T_{FP}, as a function of the frequency of the wave. If R_1 and R_2 are the reflectances of the two mirrors, one might naively guess that T_{FP} is given by the product $(1 - R_1)(1 - R_2)$. This is wrong, because the field of the transmitted wave needs to be calculated by taking into account the interference process arising by the superposition of the multiple reflections of the waves bouncing back and forth between the two mirrors.

The treatment is analogous to the one performed for the anti-reflection coating. By introducing the field reflection and transmission coefficients of the two mirrors, ρ_1, τ_1, ρ_2, τ_2, and calling L the distance between the two mirrors, the electric field of the transmitted wave, E_t, is given by a converging geometric series:

$$E_t = E_i \tau_1 \tau_2 e^{i\Delta/2}[1 + \rho_1\rho_2 e^{i\Delta} + (\rho_1\rho_2 e^{i\Delta})^2 + \cdots] = E_i \frac{\tau_1\tau_2 e^{i\Delta/2}}{1 - \rho_1\rho_2 e^{i\Delta}}, \quad (2.115)$$

where E_i is the electric field of the incident wave, and Δ is the phase shift introduced by the round-trip propagation between the two mirrors. If λ is the wavelength of the incident wave and n the refractive index of the medium that fills the cavity, $\Delta = 4\pi nL/\lambda$. Putting $\rho_1 = \sqrt{R_1}e^{i\phi_1}$, $\rho_2 = \sqrt{R_2}e^{i\phi_2}$, $|\tau_1|^2 = 1 - R_1$, and $|\tau_2|^2 = 1 - R_2$, it is found:

$$T_{FP} = \frac{|E_t|^2}{|E_i|^2} = \frac{1 + R_1R_2 - R_1 - R_2}{1 + R_1R_2 - 2\sqrt{R_1R_2}\cos\delta}, \quad (2.116)$$

where $\delta = \Delta + \phi_1 + \phi_2$. The Fabry-Perot interferometer is usually made of two identical mirrors. Putting $R_1 = R_2 = R$, (2.116) becomes:

$$T_{FP} = \frac{(1 - R)^2}{1 + R^2 - 2R\cos\delta}. \quad (2.117)$$

As shown in Fig. 2.22, T_{FP} is a periodic function of the phase delay δ, and takes the maximum value, $T_{max} = 1$, when $\cos\delta = 1$, that is $\delta = 2q\pi$, q being a positive integer. T_{FP} is minimum when $\cos\delta = -1$, that is, $\delta = (2q + 1)\pi$. The minimum value is: $T_{min} = [(1 - R)/(1 + R)]^2$. For instance, if $R = 0.99$, $T_{min} = 0.25 \times 10^{-4}$. The ratio between T_{max} and T_{min} is the contrast factor F:

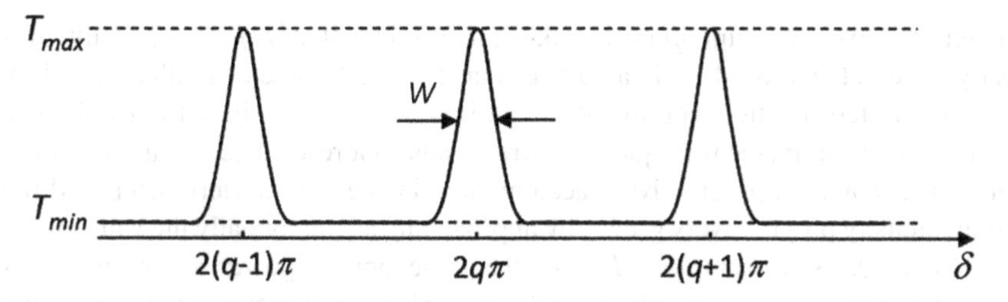

Fig. 2.22 Transmittance of the Fabry-Perot interferometer as a function of the round-trip phase shift δ, which is linearly related to the frequency of the light wave

$$F = \frac{T_{\max}}{T_{\min}} = 1 + \frac{4R}{(1-R)^2}. \tag{2.118}$$

Since δ is linearly dependent on the frequency ν of the incident wave, the abscissa scale in Fig. 2.22 can be considered as a frequency scale. As such the figure shows that the interferometer is a periodic filter in frequency-space. The frequencies corresponding to maximum transmission (resonance frequencies) are given by:

$$\nu_q = \frac{c}{2nL}(q - \frac{\phi_1 + \phi_2}{2\pi}). \tag{2.119}$$

The frequency distance between two consecutive transmission maxima, called free spectral range, is:

$$\Delta\nu = \nu_{q+1} - \nu_q = \frac{c}{2nL}. \tag{2.120}$$

An important parameter characterizing the performance of the interferometer is the width at half-height, W, of the transmission peak, expressed in units of radians. If R is close to 1, the approximate expression of W derived from (2.117) is: $W \simeq 4/\sqrt{F} \simeq 2(1-R)$. In frequency units, the peak width, $\delta\nu$, is derived by noting that $\delta\nu/\Delta\nu = W/(2\pi)$. Therefore:

$$\delta\nu = \frac{1-R}{\pi}\Delta\nu \tag{2.121}$$

As an example, if $R = 0.99$, $\delta\nu \approx \Delta\nu/300$. Taking $L = 1$ cm and $n = 1$, one finds $\Delta\nu = 15$ GHz and $\delta\nu = 50$ MHz.

Equation (2.121) seems to suggest that, once fixed $\Delta\nu$, the resolving power of the interferometer could be made arbitrarily large by choosing a value of R very close to 1. However, when $1 - R$ is made very small, there are effects due to diffraction or misalignment losses that become important. In practice, it is difficult to obtain values of the ratio $\Delta\nu/\delta\nu$ (usually called finesse of the Fabry-Perot interferometer) larger than 300. The minimum achievable frequency resolution is of the order of 1 MHz.

The finding that a sequence of two highly reflecting mirrors could give rise to a complete transmission of the beam power, when $\cos\delta = 1$, may appear rather counter-intuitive. It is important to take into account that T_{FP} is the result of a steady-state calculation. If it is assumed that the interferometer is illuminated by an intensity step-function, starting at $t = 0$, it is clear that the initial transmittance is very low. If the incident frequency corresponds to a resonance, what happens is that optical power is progressively accumulated inside the interferometer, and the output intensity progressively grows, by approaching asymptotically the value of the input intensity. As an example, if $R = 0.99$ and the incident power is 1 W, the output power of 1 W will be reached only when the power bouncing back and forth inside the cavity attains the value of 100 W. It can be shown that the time constant controlling such a transient is the reciprocal of the frequency resolution $\delta\nu$. In other terms, the larger is the resolving power of the instrument, the longer is its time-response.

In order to use the Fabry-Perot interferometer for spectral analysis, one should be able to continuously shift the transmission peaks by changing the optical path nL, so that a scan of the frequency interval between two consecutive resonant frequencies can be performed. What is the change in L, δL, required to scan a given free spectral range? Assuming that the interferometer is initially tuned to ν_q, the condition to satisfy is that the new resonance ν'_{q-1} corresponding to the modified length $L + \delta L$ should coincide with ν_q:

$$\nu_q = q\frac{c}{2L} = \nu'_{q-1} = (q-1)\frac{c}{2(L-\delta L)} \tag{2.122}$$

From (2.122) it is found:

$$\delta L = \frac{L}{q} = \frac{c}{2\nu_q} = \frac{\lambda_q}{2}. \tag{2.123}$$

Therefore in order to scan a free spectral range it is sufficient to change the mirror distance by half wavelength.

In practical cases, the scan is performed by attaching one of the mirrors to a piezoceramic, and varying the ceramic thickness by the application of a slow voltage ramp. The mirror displacement is proportional to the applied voltage, and therefore changes linearly with time. In such a measurement, the time scale is transformed into a frequency scale, and so the observed temporal dependence of the transmitted power coincides with the spectral distribution of the incoming optical signal.

As an alternative approach, scanning of the resonant frequency can also be performed by keeping L fixed and changing the refractive index of the medium inside the Fabry-Perot cavity. This can be done by putting the interferometer inside a sealed gas cell and sweeping the pressure.

When L is larger than a few centimeters, spherical mirrors are used instead of plane mirrors. The advantage of this approach is that the alignment tolerance is less critical, and diffraction losses are lower.

To avoid ambiguity in spectral analysis, the spectral width of the incoming beam should be narrower than the free spectral range. This means that the Fabry-Perot interferometer is useful only for narrow-band signals.

2.5 Waves in Anisotropic Media

Manipulation of light polarization is a key issue in several applications. Main goals include controlled modification of the polarization state of a beam or selective transmission of that portion of beam polarized in a particular linear direction. To this end, components based on materials that exhibit an anisotropy in transmission or refraction are usually exploited. All media constituted by randomly positioned atoms (or molecules), like gases, liquids, and amorphous solids (glasses) are isotropic, that is, their macroscopic properties are independent of the orientation with respect to an applied field. In contrast, ordered media (crystals) or partially ordered media (liquid crystals) may be anisotropic. A consequence of this anisotropy is that the velocity of an optical wave inside the medium will depend on its propagation direction and polarization state. As will be seen, anisotropic materials can be used to change the polarization state of an optical beam.

Inside an anisotropic medium the relation between the electric field and the electric displacement is described by a tensor:

$$D_i(\mathbf{r}, \omega) = \varepsilon_o \sum \varepsilon_{ij} E_j(\mathbf{r}, \omega). \tag{2.124}$$

The indices i, j can take values 1, 2, 3, corresponding to the axes x, y, z. Equation (2.124) describes a system of three linear equations connecting the Cartesian components of $\mathbf{E}$ to those of $\mathbf{D}$. The electric permittivity tensor, ε_{ij}, is expressed by a 3×3 matrix. It can be demonstrated that such a matrix is symmetric, which means that $\varepsilon_{ij} = \varepsilon_{ji}$. Symmetric matrices can be made diagonal by an appropriate choice of the coordinate system. With reference to the diagonal axes, usually called principal axes, the off-diagonal elements of the electric permittivity matrix vanish:

$$\begin{pmatrix} \varepsilon_{11} & 0 & 0 \\ 0 & \varepsilon_{22} & 0 \\ 0 & 0 & \varepsilon_{33} \end{pmatrix} \tag{2.125}$$

From now on, unless otherwise specified, it will be assumed that the reference system is the system of principal axes.

In the general case in which $\mathbf{E}$ has an arbitrary direction inside the anisotropic medium, the vector $\mathbf{D}$ will not be parallel to $\mathbf{E}$. Considering a monochromatic plane wave with wave vector $\mathbf{k}$, propagating inside an anisotropic medium, Maxwell's equations indicate that $\mathbf{D}$ and $\mathbf{H}$ are both perpendicular to $\mathbf{k}$, whereas $\mathbf{E}$ is perpendicular to $\mathbf{H}$, but not, in general, to $\mathbf{k}$. However, if $\mathbf{E}$ is directed along one of the principal axes, $\mathbf{E}$ and $\mathbf{D}$ are parallel, but the relative permittivity will take different values depending on the specific principal axis. Note that the polarization direction inside an anisotropic medium is the direction of $\mathbf{D}$, not of $\mathbf{E}$.

It is useful to distinguish three situations:

- isotropic medium, $\varepsilon_{11} = \varepsilon_{22} = \varepsilon_{33}$, the refractive index $n = \sqrt{\varepsilon_{11}}$ is a scalar quantity. The wave propagation velocity is the same for all directions and all polarizations.
- uniaxial medium, $\varepsilon_{11} = \varepsilon_{22} \neq \varepsilon_{33}$, it is possible to define an ordinary refractive index $n_o = \sqrt{\varepsilon_{11}} = \sqrt{\varepsilon_{22}}$ and an extraordinary refractive index $n_e = \sqrt{\varepsilon_{33}}$. The z axis is called optical axis.
- biaxial medium, $\varepsilon_{11} \neq \varepsilon_{22} \neq \varepsilon_{33}$, the medium is characterized by three distinct refractive indices.

In order to describe wave propagation in a uniaxial medium, it is better to start with a simple situation, namely that of a wave propagating in a direction perpendicular to the optical axis. Assuming that **k** is parallel to the x axis, **D** lies in the y-z plane. If the wave is linearly polarized along y, then its velocity will be c/n_o, where $n_o = \sqrt{\varepsilon_{22}}$. Analogously, if the wave is polarized along z, then the propagation velocity will be c/n_e, where $n_e = \sqrt{\varepsilon_{33}}$. In both cases, the wave polarization does not change during propagation. What happens now if the linear polarization direction is not coincident with a principal axis? By exploiting the linearity of Maxwell equations, the plane wave can be described as a linear combination of two waves, one polarized along y and the other polarized along z. The two waves travel with different velocities, and so a phase delay arises. When recombined, the resulting wave is elliptically polarized.

In the general case, in which **k** is not directed along a principal axis, the situation is more complicated. It would be useful to discover if it is also the case that there are two "special" linear polarization directions possessing the property that the polarization remains unchanged during propagation. Luckily, straightforward theory shows not only that those special directions exist, but also provides the means of determining them. Once the directions are found, the propagation is treated by describing **D** as the sum of two principal polarization modes.

The principal directions are found by using a geometric construction based on the index ellipsoid that is described by the equation:

$$\frac{x^2}{{n_1}^2} + \frac{y^2}{{n_2}^2} + \frac{z^2}{{n_3}^2} = 1. \tag{2.126}$$

Consider the ellipse generated by intersecting the ellipsoid with the plane perpendicular to **k** and containing the origin of the coordinate system, as shown in Fig. 2.23. The two principal modes have polarization directions given by the ellipse axes. In addition, if n_a and n_b are the lengths of the ellipse semi-axes, then the propagation velocities of the two modes are c/n_a and c/n_b, respectively.

In the case of a uniaxial medium, by taking the symmetry axis parallel to z, and putting $n_1 = n_2 = n_o$, and $n_3 = n_e$, (2.126) becomes:

$$\frac{x^2 + y^2}{{n_o}^2} + \frac{z^2}{{n_e}^2} = 1. \tag{2.127}$$

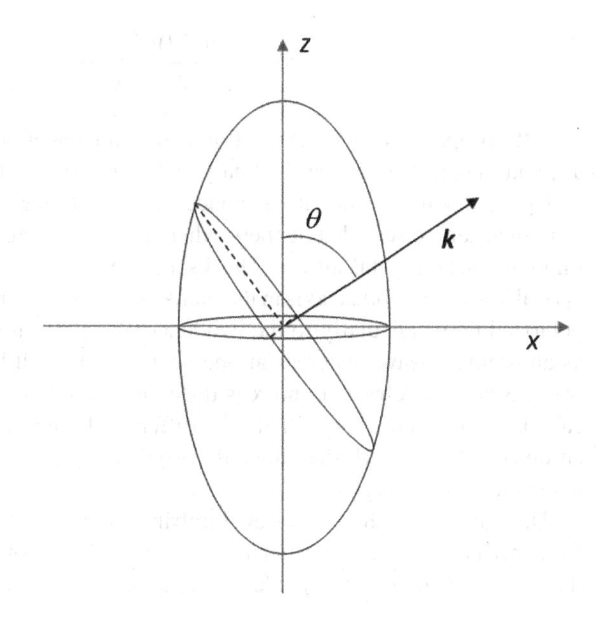

Fig. 2.23 Index ellipsoid

From now on, the discussion is restricted to uniaxial media. It can be assumed without loss of generality that **k** is in the plane x-z, forming an angle θ with the z axis. The plane that is perpendicular to **k** and passes through the origin has equation: $x\sin\theta + z\cos\theta = 0$. The intersection of this plane with the index ellipsoid gives an ellipse that has one axis coincident with y, and the other one, placed in the x-z plane, forming an angle θ with z. The length of the first semi-axis is n_o. The length of the second semi-axis, called n_θ, satisfies the equation:

$$\frac{1}{n_\theta^2} = \frac{\cos^2\theta}{{n_o}^2} + \frac{\sin^2\theta}{{n_e}^2} \tag{2.128}$$

A generic plane wave with wave vector **k** can be described as the superposition of an ordinary wave and an extraordinary wave. The former, polarized perpendicularly to the plane defined by **k** and z, travels at velocity c/n_o. The latter, polarized in the plane defined by **k** and z, travels at velocity c/n_θ. If $\theta = 0$, then $n_\theta = n_o$, that is, for a wave propagating along the optical axis the medium behaves as though it were isotropic. If $\theta = 90°$, the extraordinary wave is polarized along z, and has a propagation velocity c/n_e.

For the ordinary wave, **D** has no components along z, therefore **D** is parallel to **E**, and the Poynting vector is parallel to **k**. For the extraordinary wave, if $0 < \theta < 90°$, **D** has nonzero components along both z and x. Therefore **D** is not parallel to **E**. The consequence is that the Poynting vector forms with **k** a non-zero angle, ψ, called walk-off angle. ψ is given by the relation:

$$\tan\psi = \frac{\sin(2\theta)n_\theta^2}{2}\left(\frac{1}{n_e{}^2} - \frac{1}{n_o{}^2}\right) \qquad (2.129)$$

If the propagating wave is a collimated beam, the effect of walk-off is to generate a spatial separation between ordinary and extraordinary beams.

Up to now the treatment has concerned the propagation inside the crystal. It is also important to see what happens when a wave coming from an isotropic medium, impinges on the crystal surface. The discussion is here limited to the case of a uniaxial crystal having the optical axis in the incidence plane. An incident wave that is linearly polarized perpendicularly to the incidence plane (σ case) propagates in the crystal as an ordinary wave, whereas an incident π wave will behave as an extraordinary wave. Since the refractive index is different in the two cases, the refraction angles calculated by Snell's law will also be different. In the case of arbitrary polarization, an object seen through the uniaxial crystal will present a double image, hence the name birefringent crystal.

The refractive indices of a few birefringent crystals are given in Table 2.3. Birefringence is said to be positive if $n_e > n_o$, as in the case of quartz crystal, and negative if $n_e < n_o$, as in the case of calcite crystal.

2.5.1 Polarizers and Birefringent Plates

Many applications require a well-defined polarization state. Conventional optical sources produce light having a polarization state that changes randomly with time (unpolarized light). Some lasers emit polarized light. Birefringent crystals are used to generate polarized light or to modify the state of polarization of a light beam.

The simplest polarizer is made from a dichroic material. This is an anisotropic material having, for instance, a large absorption coefficient for light polarized along the optical axis, and a small absorption for light polarized perpendicularly to the

Table 2.3 Refractive indices of some birefringent crystals

Crystal	n_o	n_e	λ (nm)
Calcite, $CaCO_3$	1.658	1.486	589
Quartz, SiO_2	1.543	1.552	589
Rutile, TiO_2	2.616	2.903	589
Corundum, Al_2O_3	1.768	1.760	589
Lithium niobate, $LiNbO_3$	2.304	2.212	589
Barium borate, BaB_2O_4	1.678	1.553	589

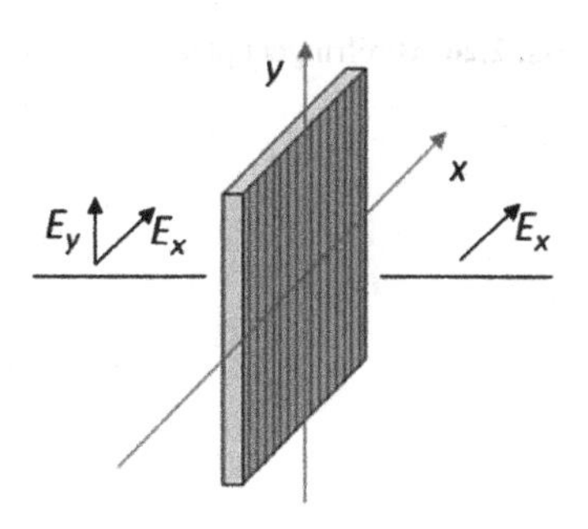

Fig. 2.24 Polaroid polarizer. The polymer chains are parallel to y, and so only the linear polarization parallel to x is transmitted

optical axis. The most common dichroic material, Polaroid, is composed of a parallel arrangement of linear chains of conjugated polymers (see Fig. 2.24).

The optical axis coincides with the chain direction. The polymer presents high electric conductivity along the chain and low conductivity across the chain. In a grossly simplified description it behaves as a conductor if the applied electric field is parallel to the optical axis, and as an insulator if the electric field is perpendicular to the optical axis. In practice the component polarized perpendicularly to the chains also suffers a non-negligible attenuation. Polarizers made of Polaroid are cheap, but the selectivity is not very large (a typical value of the extinction ratio, defined as the transmittance ratio between unwanted and wanted polarization state, could be $\approx 1/500$). Since absorption heats the material, high-intensity beams damage the polarizer.

High-selectivity polarizers, having an extinction ratio about 10^5, consist of a pair of prisms made of a birefringent crystal, typically calcite. One of the most utilized polarizers is the Glan–Thompson polarizer, shown in Fig. 2.25, which is made of two calcite prisms cemented together by their long faces. The cement has a refractive index intermediate between the ordinary and the extraordinary indices of calcite. The optical axes of the calcite crystals are parallel and aligned perpendicular to the incidence plane, that is, the plane of the figure. Light of arbitrary polarization is split into two rays, experiencing different refractive indices; the π-polarized ordinary ray is totally internally reflected from the calcite-cement interface, leaving the σ-polarized extraordinary ray to be transmitted. The prism can therefore be used as a polarizing beam splitter. Traditionally a resin called Canada balsam was used as the cement in assembling these prisms, but this has largely been replaced by synthetic polymers.

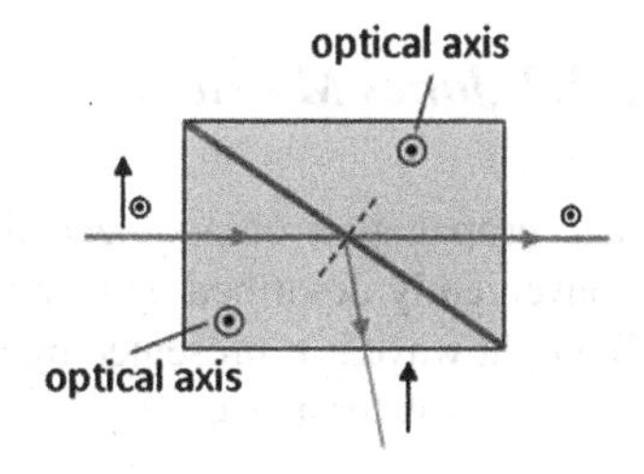

Fig. 2.25 Glan-Thompson prism

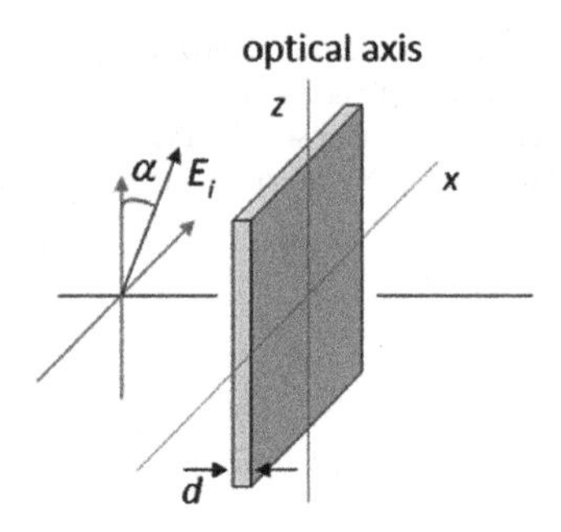

Fig. 2.26 Birefringent plate

The Glan-Thompson polarizer operates over a very wide wavelength range, from 350 nm to 2.3 µm. This type of polarizer can handle much higher optical powers than the dichroic polarizers.

Thin plates made of a birefringent crystal are used to change the polarization state of a light beam. The plate faces are parallel to the x-z plane, and the light beam travels in the y direction, perpendicular to the optical axis z, as shown in Fig. 2.26. If the incident beam is linearly polarized in the x direction (ordinary wave), then the propagation across the plate does not modify the polarization state, but simply introduces a phase shift $\Delta\phi_o = 2\pi n_o d/\lambda$, where d is the plate thickness. Similarly, no modification of the polarization state occurs for the beam polarized along z, which is phase-shifted by: $\Delta\phi_e = 2\pi n_e d/\lambda$. An input wave linearly polarized in an arbitrary direction can be described as the sum of an ordinary and an extraordinary wave. The two waves are initially in phase, but acquire, on crossing the plate, a phase difference $\Delta\phi = 2\pi(n_e - n_o)d/\lambda$. In general, the output polarization is elliptical. An important particular case is a birefringent plate that gives a phase difference of $|\Delta\phi| = \pi$. This is known as a half-wave plate. When linearly-polarized light is incident upon a half-wave plate, the emergent light is also linearly-polarized. If the polarization direction of the incident light makes an angle α with the optical axis, then the polarization direction of the emergent beam makes an angle $-\alpha$ with the same axis, therefore the direction of polarization has been rotated through an angle 2α, as shown in Fig. 2.27. A plate for which $|\Delta\phi| = \pi/2$ is called a quarter-wave plate. When linearly-polarized light is incident upon a quarter-wave plate, the emergent light is in general elliptically polarized. If, however, $\alpha = \pi/4$, the emergent light is circularly polarized.

2.5.2 *Jones Matrices*

The propagation of polarized light beams trough anisotropic optical elements is conveniently described by using the Jones formalism, in which the electric field of the light wave is represented by a two-component vector and the optical element by a 2×2 transfer matrix.

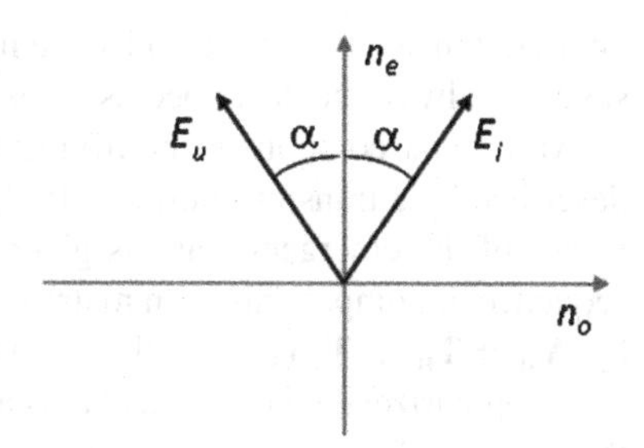

Fig. 2.27 Rotation of the polarization direction in a half-wave plate

In general, the electric field of a monochromatic plane wave propagating along z can be described as the sum of two components:

$$\mathbf{E}(z,t) = E_x \exp[-i(\omega t - kz + \phi_x)]\mathbf{x} + E_y \exp[-i(\omega t - kz + \phi_y)]\mathbf{y}, \quad (2.130)$$

where $\mathbf{x}$ and $\mathbf{y}$ are unit vectors, directed along x and y, respectively. Equation (2.130) can be written as:

$$\mathbf{E}(z,t) = E_o \exp[-i(\omega t - kz + \phi_x)]\mathbf{V}, \quad (2.131)$$

where $E_o = \sqrt{E_x{}^2 + E_y{}^2}$, and $\mathbf{V}$ is the Jones vector represented by the column matrix:

$$\begin{pmatrix} \cos\alpha \\ e^{i\phi} \sin\alpha \end{pmatrix}, \quad (2.132)$$

where $\alpha = \arctan(E_y/E_x)$ and $\phi = \phi_y - \phi_x$.

The adopted sign convention considers the angles measured counter-clockwise as being positive.

The vector (2.132) corresponds to a generic elliptical polarization. If the wave is linearly polarized, $\phi = 0$. The following Jones vectors describe a linearly polarized field with a polarization direction forming the angle α with the x axis, or directed along the x axis, or along the y axis, respectively:

$$\begin{pmatrix} \cos\alpha \\ \sin\alpha \end{pmatrix}; \begin{pmatrix} 1 \\ 0 \end{pmatrix}; \begin{pmatrix} 0 \\ 1 \end{pmatrix}. \quad (2.133)$$

For a circularly polarized wave, $\alpha = \pi/4$ and $\phi = \pm\pi/2$. Therefore the Jones vectors associated with clockwise and counter-clockwise circular polarization are respectively given by:

$$\frac{1}{\sqrt{2}}\begin{pmatrix} 1 \\ i \end{pmatrix}; \frac{1}{\sqrt{2}}\begin{pmatrix} 1 \\ -i \end{pmatrix}. \quad (2.134)$$

Two waves represented by the vectors $\mathbf{V_1}$ and $\mathbf{V_2}$ have orthogonal polarization states if the scalar product of the two vectors is zero. As an example, the linear polarization states along x and y are mutually orthogonal. An arbitrary polarization state can always be described as the sum of two orthogonal polarization states.

An optical component modifying the polarization state of an incident beam is described by a transfer matrix $\mathbf{T}$. If $\mathbf{V_i}$ represents the incident wave, then the Jones vector of the emergent wave is given by: $\mathbf{V_u} = \mathbf{T}\mathbf{V_i}$. If the beam goes through a sequence of components with matrices $\mathbf{T_1}, \ldots, \mathbf{T_n}$, then the emergent wave is given by: $\mathbf{V_u} = \mathbf{T_n} \ldots \mathbf{T_1}\mathbf{V_i}$. Note that, in general, the $\mathbf{T}$ matrices do not commute.

The polarizers having optical axis along x and y are represented, respectively, by the matrices $\mathbf{T_{px}}$ e $\mathbf{T_{py}}$:

$$\mathbf{T_{px}} = \begin{pmatrix} 1 & 0 \\ 0 & 0 \end{pmatrix}; \quad \mathbf{T_{py}} = \begin{pmatrix} 0 & 0 \\ 0 & 1 \end{pmatrix}. \tag{2.135}$$

The birefringent plate having optical axis directed along x introduces a phase shift $\Delta\phi$ between the x and y components of the electric field, without modifying the amplitude ratio E_x/E_y. It is represented by:

$$\mathbf{T_r} = \begin{pmatrix} 1 & 0 \\ 0 & \exp(i\Delta\phi) \end{pmatrix}. \tag{2.136}$$

The phase shift is given by $\Delta\phi = 2\pi(n_e - n_o)d/\lambda$, where d is the plate thickness. The half-wave plate corresponds to $\Delta\phi = \pi$, and the quarter-wave plate to $\Delta\phi = \pi/2$.

The polarization rotation matrix,

$$\mathbf{R}(\theta) = \begin{pmatrix} \cos\theta & -\sin\theta \\ \sin\theta & \cos\theta \end{pmatrix}, \tag{2.137}$$

represents an optical component that rotates the polarization direction. If the incident wave is linearly polarized at an angle θ_1, the emergent wave is linearly polarized at the angle $\theta_2 = \theta_1 + \theta$.

It is important to take into account that Jones vectors and matrices change when the coordinate system is changed. If a field is described by the vector $\mathbf{V}$, once the coordinate system is rotated by the angle θ, the same field is described by the new vector $\mathbf{V}'$:

$$\mathbf{V}' = \mathbf{R}(-\theta)\mathbf{V}. \tag{2.138}$$

A polarizer having its optical axis that forms an angle θ with the x axis is represented by the matrix:

$$\mathbf{T_{p\theta}} = \mathbf{R}(\theta)\mathbf{T_{px}}\mathbf{R}(-\theta) = \begin{pmatrix} \cos^2\theta & \sin\theta\cos\theta \\ \sin\theta\cos\theta & \sin^2\theta \end{pmatrix}. \tag{2.139}$$

Similarly, the matrix describing an optical retarder (birefringent plate) having the optical axis rotated by θ with respect to x is written as:

$$\mathbf{T}_{\mathbf{r}\theta} = \mathbf{R}(\theta)\mathbf{T}_{\mathbf{r}}\mathbf{R}(-\theta)$$
$$= \begin{pmatrix} 1 + [\exp(i\Delta\phi) - 1]\sin^2\theta & [1 - \exp(i\Delta\phi)]\sin\theta\cos\theta \\ [1 - \exp(i\Delta\phi)]\sin\theta\cos\theta & \exp(i\Delta\phi) - [\exp(i\Delta\phi) - 1]\sin^2\theta \end{pmatrix}. \quad (2.140)$$

2.5.3 Rotatory Power

Certain materials present a particular type of birefringence: the two fundamental modes of propagation are circularly rather than linearly polarized waves, with right- and left-circular polarizations traveling at different velocities, c/n^+ and c/n^-. This property, known as optical activity, is typical of materials having an intrinsically helical structure at the molecular or crystalline level.

As mentioned in Sect. 1.3, a linearly polarized wave can be decomposed into the sum of two circularly polarized waves having the same amplitude. By using Jones matrices, one can write:

$$\begin{pmatrix} \cos\theta \\ \sin\theta \end{pmatrix} = \frac{1}{2}\exp(-i\theta)\begin{pmatrix} 1 \\ i \end{pmatrix} + \frac{1}{2}\exp(i\theta)\begin{pmatrix} 1 \\ -i \end{pmatrix}. \quad (2.141)$$

If L is the thickness of the optically active medium, the phase delay between the two circularly polarized waves is given by: $\phi = 2\pi L(n^+ - n^-)/\lambda$. The recombination of the two emergent waves produces a linearly polarized beam (see Fig. 2.28) with a polarization direction rotated with respect to that of the incident beam:

$$\frac{1}{2}\exp(-i\theta)\begin{pmatrix} 1 \\ i \end{pmatrix} + \frac{1}{2}\exp[i(\theta - \phi)]\begin{pmatrix} 1 \\ -i \end{pmatrix}$$
$$= \exp(-i\phi/2)\begin{pmatrix} \cos(\theta - \phi/2) \\ \sin(\theta - \phi/2) \end{pmatrix}. \quad (2.142)$$

The rotation per unit length, called rotatory power ρ, is:

$$\rho = \frac{\phi}{2L} = \frac{\pi(n^+ - n^-)}{\lambda}. \quad (2.143)$$

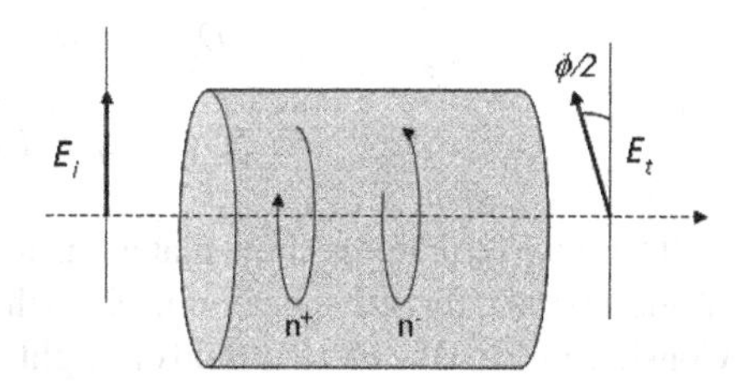

Fig. 2.28 Rotatory power

A typical optically active crystal is quartz. For propagation along the optical axis, the rotatory power is 4 rad/cm at $\lambda = 589$ nm. Also disordered media containing chiral molecules, as, for instance, sugar molecules, can show optical activity. For an aqueous solution containing 0.1 g/cm^3 of sugar, $\rho = 10^{-2}$ rad/cm. The sugar concentration can therefore be determined by measuring the rotatory power.

From a formal point of view, the optical activity can be explained only assuming the presence of anti-symmetric off-diagonal terms inside the ε_{ij} matrix. Such a presence would contradict the assertion that the matrix ε_{ij} is symmetric. However, it can be shown that such a symmetry property applies only for spatially homogeneous fields, that is, infinite wavelength fields. If λ is finite, the local value of **D** depends, in general, on **E** and its spatial derivatives. At first order:

$$D_i = \varepsilon_o \sum_{jm} \left(\varepsilon_{ij} E_j + i \gamma_{ijm} \frac{\partial E_j}{\partial x_m} \right). \tag{2.144}$$

whereas ε_{ij} is symmetrical, the third-order tensor γ_{ijm} is anti-symmetrical with respect to the indices ij: $\gamma_{ijm} = -\gamma_{jim}$. If there is no absorption, then γ_{ijm} is real. In the case of an incident plane wave with wave vector **k**, it is found that $\partial E_j / \partial x_m = i E_j k_m$. Hence:

$$D_i = \varepsilon_o \sum_{jm} \left(\varepsilon_{ij} E_j + i \gamma_{ijm} E_j k_m \right). \tag{2.145}$$

D is related to **E** by the new electric permittivity tensor:

$$\varepsilon'_{ij} = \varepsilon_{ij} + i \sum_m \gamma_{ijm} k_m, \tag{2.146}$$

which is the sum of a symmetric and an anti-symmetric term, the latter vanishing if the modulus of **k** vanishes, that is, if the wavelength becomes infinite.

For an isotropic medium, (2.145) becomes:

$$D_i = \varepsilon_o [\varepsilon_r E_i + i \gamma (\mathbf{E} \times \mathbf{k})_i], \tag{2.147}$$

where $(\mathbf{E} \times \mathbf{k})_i$ denotes the i-th component of the cross product, and γ is a scalar quantity. Assuming **k** parallel to z, the **D** components along x and y, as derived from (2.147), are:

$$D_x = \varepsilon_o [\varepsilon_r E_x + i \gamma k E_y] \tag{2.148}$$

$$D_y = \varepsilon_o [\varepsilon_r E_y - i \gamma k E_x]. \tag{2.149}$$

These two equations show that a linear polarization cannot propagate unchanged inside an optically active medium, due to the fact that E_x generates D_y, and viceversa. Considering that the electric field of a right- (left-) circularly polarized beam is written as: $E^{\pm} = E_x \pm iE_y$, by using (2.148) and (2.149) the quantities $D^{\pm} = D_x \pm iD_y$ may be written as:

$$D^{+} = \varepsilon_o[\varepsilon_r + \gamma k]E^{+} \tag{2.150}$$

$$D^{-} = \varepsilon_o[\varepsilon_r - \gamma k]E^{-} \tag{2.151}$$

These two equations show that circular polarization indeed represents a propagation mode inside the optically active medium. The refractive indices seen by the two circular polarizations are: $n^{\pm} = \sqrt{\varepsilon_r \pm \gamma k}$. Assuming that the second term under square root is much smaller than the first one, putting $k = n\omega/c$ and $\varepsilon_r = n^2$, one obtains the approximate relation: $n^{\pm} = n[1 \pm \gamma k/(2)]$, and, consequently, $\rho = \pi\gamma k/(n\lambda)$.

2.5.4 Faraday Effect

There are various means to change the optical properties of a medium by applying an external field. In this section the magneto-optic effect known as Faraday effect is described. It consists in the induction of rotatory power through the application of a static magnetic field. The effect is longitudinal, since the applied magnetic field **B** is parallel to the propagation direction of the linearly-polarized optical wave. The polarization direction is rotated by an angle $\theta = C_V BL$, where L is the propagation length and C_V is the Verdet constant. The sign of θ changes with the sign of **B**, but it does not change by reversing the propagation direction. Therefore the rotation effect is doubled if the beam travels back and forth inside the medium.

In the standard configuration of the Faraday rotator the medium has a cylindrical shape and is placed inside a solenoid. The longitudinal magnetic field is created by an electric current flowing into the solenoid.

In the presence of the magnetic field, the relation between **D** and **E** can be written, similarly to (2.144), as:

$$\mathbf{D} = \varepsilon_o[\varepsilon_r\mathbf{E} + i\gamma_B(\mathbf{B} \times \mathbf{E})], \tag{2.152}$$

where γ_B is called magneto-gyration coefficient. The rotatory power is given by:

$$\rho = C_V B = -\frac{\pi\gamma_B B}{n\lambda}. \tag{2.153}$$

The big difference between a passive rotator, such as a quartz plate, and the Faraday rotator is made clear by considering the situation in which the wave transmitted in one direction is reflected back and retransmitted in the opposite direction. In the case of the passive rotator there is no net polarization rotation because the rotation generated in the forward path is canceled during the backward path, while in the case of the Faraday rotator the effect is doubled. The Faraday rotator is said to be a nonreciprocal device.

The Faraday effect can be found in gases, liquids, and solids. Since its microscopic origin is connected to the Zeeman effect, the Verdet constant becomes larger when the wavelength of the optical wave gets close to absorption lines. The most commonly utilized materials are doped glasses and crystals, like YIG (yttrium iron garnet), TGG (terbium gallium garnet) and TbAlG (terbium aluminum garnet). As an example, the Verdet constant of TGG is: $C_V = -134$ rad/(T · m) at $\lambda = 633$ nm, and becomes -40 rad/(T · m) at $\lambda = 1064$ nm.

There are some ferromagnetic materials (containing Fe, Ni, Co) that exhibit large values of C_V, but they also have a very large absorption coefficient, so the Faraday rotation can only be observed in reflection. They have an important application in optical memories, as will be mentioned in Chap. 8.

2.5.5 Optical Isolator

An optical isolator is a device that transmits light in only one direction, thereby preventing reflected light from returning back to the source. Isolators are utilized in many laser applications, since back-reflected beams often have deleterious effects on the operating conditions of optical devices.

A Faraday isolator may be constructed by placing a Faraday rotator between two polarizers, as shown in Fig. 2.29. The optical axis of polarizer 1 is along x, the Faraday cell rotates clockwise the polarization direction by $\pi/4$, the optical axis of polarizer 2 makes a $\pi/4$ angle with the x axis, so that the forward traveling wave is fully transmitted by the device. If a portion of the transmitted wave is back-reflected by a

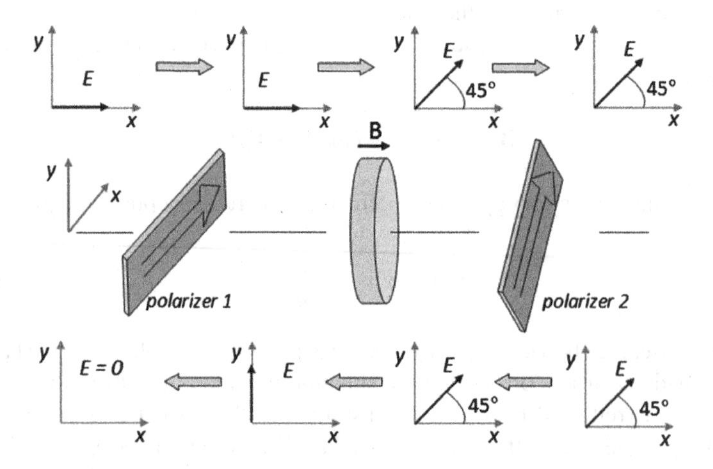

Fig. 2.29 Faraday isolator

subsequent optical component, it travels inside the isolator in the backward direction. Assuming that reflection does not alter the polarization state, the reflected beam is fully transmitted by polarizer 2, and its polarization direction suffers a clockwise rotation of $\pi/4$ by the Faraday cell. The backward beam becomes y-polarized, and therefore is blocked by polarizer 1.

The behavior of the isolator can be mathematically described by Jones matrices. Let $\mathbf{V_i}$ and $\mathbf{V_u}$ be the Jones vectors of the incident and emergent waves, respectively, and call $\mathbf{T_1}$, $\mathbf{T_2}$, $\mathbf{T_3}$, the matrices representing polarizer 1, Faraday rotator, and polarizer 2, respectively. It is important to recall that all vectors and matrices must be expressed with reference to the same x-y coordinate system. The output field is related to the input field by the following expression:

$$\begin{aligned}\mathbf{V_u} &= \mathbf{T_3T_2T_1V_i} \\ &= \begin{pmatrix}1/2 & 1/2\\ 1/2 & 1/2\end{pmatrix}\begin{pmatrix}\sqrt{2}/2 & -\sqrt{2}/2\\ \sqrt{2}/2 & \sqrt{2}/2\end{pmatrix}\begin{pmatrix}1 & 0\\ 0 & 0\end{pmatrix}\begin{pmatrix}1\\ 0\end{pmatrix} = \begin{pmatrix}\sqrt{2}/2\\ \sqrt{2}/2\end{pmatrix}. \end{aligned} \tag{2.154}$$

As expected, the emergent wave is linearly polarized at $\pi/4$ with respect to the x axis. The back-reflected beam crosses the sequence of devices in opposite order, and is completely blocked by the isolator:

$$\mathbf{V'_u} = \mathbf{T_1T_2T_3V_u} = \begin{pmatrix}0\\ 0\end{pmatrix}. \tag{2.155}$$

Note that $\mathbf{T_3}$ seems to have no role in the proposed scheme. In practice, however, it may happen that back-reflection modifies the polarization state. In such a case the role of polarizer 2 becomes important.

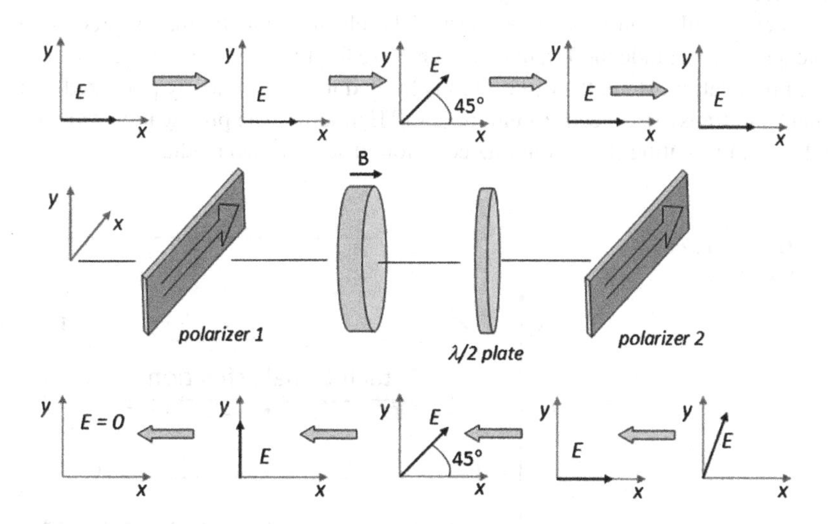

Fig. 2.30 Faraday isolator including a half-wave plate

The sequence of a Faraday rotator plus a half-wave plate constitutes another nonreciprocal device, as schematically depicted in Fig. 2.30. If the former rotates the polarization direction by $\pi/4$, and the latter by $-\pi/4$, a linearly polarized beam does not change its polarization state by crossing the device. On the contrary, for a beam coming from the opposite direction, the two rotations have the same sign, so that its polarization direction is rotated by $\pi/2$. Such a device can be used to make uni-directional the wave propagation in a circular path.

2.6 Waveguides

Waveguides are strips of dielectric material embedded in another dielectric material of lower index of refraction. Starting from the advent of optical communications, the development of integrated optical systems has become more and more important. In such systems the light propagates in a confined regime through optical waveguides, instead of free space, without suffering from the effect of diffraction.

A light beam traveling in a direction that is only slightly tilted with respect to the guide axis is totally reflected at the boundary between the two media, provided the incidence angle is larger than the limit angle. So that the guide can completely trap a light beam in a diffraction-less propagation: the optical power should be mainly confined inside the guide, with only an evanescent tail extending outside.

Many aspects of guided propagation can be understood by studying the simple case of a planar dielectric waveguide, consisting of a slab of thickness d and refractive index n_1, usually called core, surrounded by a cladding of lower refractive index n_2, as shown in Fig. 2.31. In a planar waveguide, confinement is guaranteed only in one dimension. Light rays making angles θ with the z-axis, in the x-z plane, undergo total internal reflection if $\cos\theta \geq n_2/n_1$. The electromagnetic theory predicts the existence of waveguide modes, that is, transverse field distributions that are invariant in the propagation. Here, the discussion is limited to waves linearly polarized along x, that is, to transverse electric-field modes (TE modes). The propagation properties are derived by writing the Helmholtz equation inside the two media,

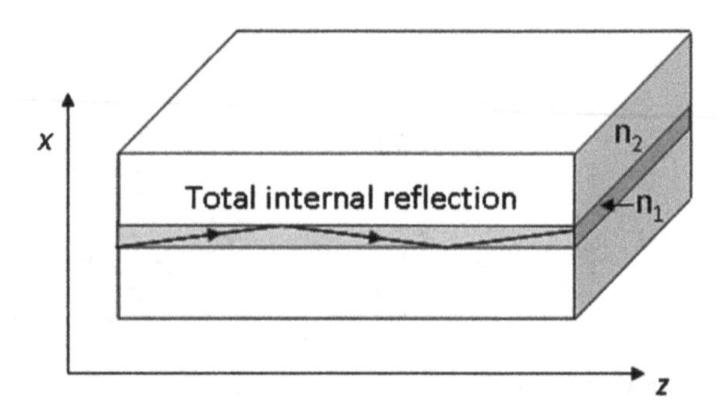

Fig. 2.31 Scheme of a planar waveguide

$$(\nabla^2 + (n_1 k_o)^2)\mathbf{E}(\mathbf{r}) = 0 \tag{2.156}$$

$$(\nabla^2 + (n_2 k_o)^2)\mathbf{E}(\mathbf{r}) = 0, \tag{2.157}$$

and imposing the boundary conditions.

By assuming that the wave propagates along z with a propagation constant β, the electric field can be generally written as:

$$E(x, z, t) = A(x)\exp[i(\beta z - \omega t)] \tag{2.158}$$

By substitution of (2.158) into (2.156), the equation relative to medium 1 (the core) becomes:

$$\frac{d^2 A}{dx^2} + [(n_1 k_o)^2 - \beta^2]A(x) = 0. \tag{2.159}$$

Similarly, inside the cladding:

$$\frac{d^2 A}{dx^2} + [(n_2 k_o)^2 - \beta^2]A(x) = 0 \tag{2.160}$$

Since the propagation constant of the guided wave must be intermediate between $n_2 k_o$ and $n_1 k_o$, the solution of (2.159) is a periodic function, $A_1 \cos(\gamma_1 x)$, and that of (2.160) is an exponential function, $A_2\exp(-\gamma_2 x)$, where the real quantities γ_1 and γ_2 are given, respectively, by:

$$\gamma_1 = \sqrt{n_1{}^2 k_o{}^2 - \beta^2}, \tag{2.161}$$

and

$$\gamma_2 = \sqrt{\beta^2 - n_2{}^2 k_o{}^2} \tag{2.162}$$

By imposing the continuity of the field amplitude and its derivative at $x = d/2$, it is found:

$$A_1 \cos(\gamma_1 d/2) = A_2 \exp(-\gamma_2 d/2), \tag{2.163}$$

and

$$A_1 \gamma_1 \sin(\gamma_1 d/2) = A_2 \gamma_2 \exp(-\gamma_2 d/2). \tag{2.164}$$

Non-zero field amplitudes A_1 and A_2 exist only if the following condition is satisfied:

$$\tan(\gamma_1 d/2) = \frac{\gamma_2}{\gamma_1}. \tag{2.165}$$

This is a transcendental equation in the variable β. A graphic solution can be found by plotting the right- and left-hand sides as functions of β.

What has been calculated is the propagation constant of the fundamental mode. It can be shown that the waveguide can support a discrete sequence of modes, whose propagation constants are derived by generalizing (2.165) as follows:

$$\tan(\gamma_1 d/2 - m\pi/2) = \frac{\gamma_2}{\gamma_1}, \tag{2.166}$$

where $m = 0, 1, \ldots$.

Whereas a solution always exists for the fundamental mode, a value of the propagation constant β_1 of the $m = 1$ mode satisfying the condition $n_2 k_o \leq \beta_1 \leq n_1 k_o$ is found only if the mode frequency ν is larger than the cut-off frequency ν_c given by:

$$\nu_c = \frac{c}{2d\sqrt{n_1{}^2 - n_2{}^2}}, \tag{2.167}$$

By increasing the mode order m, the cut-off frequency increases. Therefore, by increasing the frequency, the guide accepts more and more propagation modes. According to (2.167), to be single-mode the waveguide must be thin, and the refractive index difference $n_1 - n_2$ needs to be small. As it will be seen in the following chapters, there are various applications requiring single-mode propagation.

Concerning the field distribution of the waveguide modes, it should be noted that, by increasing m, the field penetrates deeper into the cladding. Correspondingly, the propagation constant β_m decreases with m.

Most of the considerations developed for the propagation in slab waveguides also apply for the case of channel waveguides, in which the confinement of the light beam is guaranteed on two dimensions. A very important example of channel waveguide is the optical fiber (Fig. 2.32a), which is a waveguide with a cylindrical symmetry that will be discussed in greater details in Chap. 6. Figure 2.32b–d illustrate the typical geometries of different channel waveguides fabricated on planar substrates, which constitute the basic structures of several integrated optical devices, like optical modulators (Chap. 4) and semiconductor lasers or amplifiers (Chap. 5). In the recent years, the evolution of communication systems has driven the development of optical circuits mainly based on silicon deposited on silicon oxide (silicon on insulator, SOI) in which several electronic and photonic functions are integrated on the same

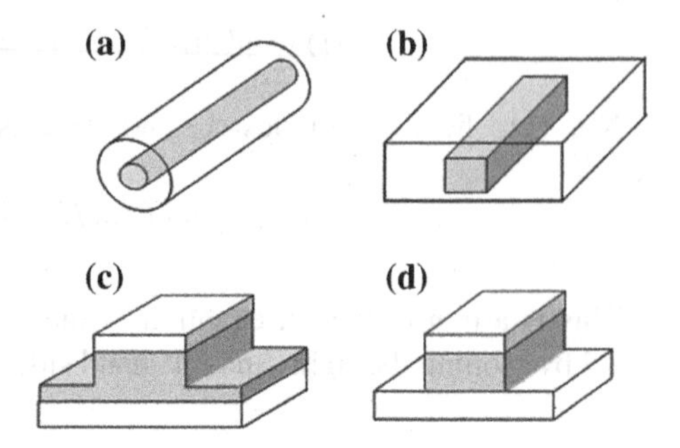

Fig. 2.32 Scheme of the most commonly used optical waveguide geometries: **a** optical fiber, **b** embedded waveguide, **c** rib waveguide, **d** ridge waveguide. The *darker regions* represent the cores in which the light is confined

chip. Miniaturization is a key issue for this application, as it will be discussed in Sect. 8.1.3, and waveguides with cross-sections of hundreds of nanometers and very high refractive contrast are needed. This provides high confinement of the light that can be bent in curved waveguides, thus making feasible the fabrication of complex networks of waveguides on a very small area.

Problems

2.1 Given that the wavelength of a light signal in vacuum is 600 nm, what will it be in a glass block having a refractive index $n = 1.50$?

2.2 A glass block having an index of 1.55 is coated with a layer of magnesium fluoride of index 1.32. For light traveling in the glass, what is the total reflection angle at the interface?

2.3 Yellow light from a sodium lamp ($\lambda = 589$ nm) traverses a tank of benzene (of index 1.50), which is 30-m long, in a time τ_1. If it takes time τ_2 for the light to pass through the same tank when filled with carbon disulfide (of index 1.63), determine the value of $\tau_2 - \tau_1$.

2.4 Light having a vacuum wavelength of 600 nm, traveling in a glass ($n = 1.48$) block, is incident at 45° on a glass-air interface. It is then totally internally reflected. Determine the distance into the air at which the amplitude of the evanescent wave has dropped to a value of 1/e of its maximum value at the interface.

2.5 A beam of light in air strikes the surface of a transparent material having an index of refraction of 1.5 at an angle with the normal of 40°. The incident light has component E-field amplitudes parallel and perpendicular to the plane-of-incidence of 10 and 20 V/m, respectively. Determine the corresponding reflected field amplitudes.

2.6 The complex index of refraction of silver at $\lambda = 532$ nm is $n' + in'' = 0.14 - 3.05i$. Calculate: (i) the reflectance of an air-silver surface at normal incidence; (ii) the transmittance of a 100 μm silver plate.

2.7 A Gaussian beam at wavelength $\lambda = 0.63$ μm having a spot size at the beam waist of $w_o = 0.5$ mm is to be focused to a spot size of 30 μm by a lens positioned at a distance of 0.5 m from the beam waist. What focal length should the lens have?

2.8 A Gaussian beam with $\lambda = 800$ nm and $w_o = 0.1$ mm at $z = 0$ is propagating along the z axis. A plano-convex lens, made of a spherical glass cap having a radius of curvature of 4 cm, is placed at $z_1 = 20$ cm. The refractive index of the glass is $n = 1.8$. Determine: (i) at which coordinate z_2 the beam is focused; (ii) the beam spot size at z_2; (iii) the lens diameter to be chosen in order to collect 99 % of the beam power.

2.9 A blazed reflection grating with pitch $d = 1200\,\text{nm}$ and blazing angle $\theta_g = 20°$ is used in the Littrow configuration. Determine which incident wavelength is diffracted at second order in the backward direction.

2.10 A Fabry-Perot interferometer made by two identical mirrors at distance $d = 1\,\text{mm}$ is illuminated by a light beam containing two wavelengths $\lambda_1 = 1064\,\text{nm}$ and $\lambda_2 = \lambda_1 + \Delta\lambda$, where $\Delta\lambda = 0.05\,\text{nm}$. Determine which is the minimum reflectance of the two mirrors required to obtain a resolving power $P_r \geq \lambda_1/\Delta\lambda$.

2.11 Consider a light beam at $\lambda = 589\,\text{nm}$ traveling along the z-axis and linearly polarized along the x-axis. The light beam crosses, in succession, a quarter-wave plate and a polarizer. The optical axis of the quarter-wave plate is in the x-y plane, forming an angle $\pi/4$ with the x-axis. The optical axis of the polarizer is parallel to the y-axis. Write the Jones matrix of the quarter-wave plate in the x-y reference frame, and calculate the fraction of the incident power that is transmitted by the polarizer.

2.12 Consider the following sequence of components: a polarizer P_1, a Faraday rotator that rotates counterclockwise the linear polarization direction of the incoming beam by 45°, a calcite half-wave plate, and a second polarizer P_2. Assume that the incoming beam travels along y and is linearly polarized along x, that the two polarizers have optical axes parallel to x, and that the optical axis of the calcite plate lies in the x-z plane. (i) Calculate which angle the optical axis of the plate should form with the x axis in order to ensure a unitary transmission. (ii) Write the Jones matrix of the half-wave plate in the x-z reference frame.

Chapter 3
The Laser

Abstract All the light sources utilized before the invention of the laser are based on spontaneous emission processes from excited atoms. Since each atom emits in a random direction with a random phase, conventional light sources are intrinsically chaotic. The laser is a completely different light source, because it is based on stimulated emission instead of spontaneous emission, which means that atoms emit in a cooperative way, generating a monochromatic and collimated light beam. The fundamental ingredient for making a laser is the optical amplifier. A mechanism selecting wavelength and direction is provided by the positive feedback that transforms the amplifier into an oscillator. In this chapter, after shortly mentioning conventional light sources, especially thermal sources, and after giving a brief historical introduction to the laser, the general scheme of the oscillator is first discussed. Successively, by using an elementary treatment of the interaction between light and atoms, the conditions under which a set of atoms can become an optical amplifier are derived. After discussing the general laser scheme, the steady-state and pulsed behavior is presented, together with a concise description of a few solid-state and gas lasers. The chapter ends with a comparison between the laser and the conventional light source.

3.1 Conventional Light Sources

Conventional light sources, like flames, incandescent lamps, and electric discharge lamps, operate by converting an input energy (chemical, thermal, and electrical, respectively) to light. The light is emitted over the full solid angle, usually with a very broad frequency spectrum, except for the case of low gas-pressure electric discharge lamps. The main application of these sources is to provide illumination, in substitution to the Sun.

It is of historical importance to discuss the emission of incandescent sources, such as the Sun (which converts nuclear energy from fusion reactions into light), or a common household tungsten-filament lamp. The study of the emission spectrum from these incandescent sources and of its dependence on the absolute temperature T has had a profound impact on modern physics. In the theoretical treatment proposed during the nineteenth century, an ideal body consisting of a closed cavity kept in

V. Degiorgio and I. Cristiani, *Photonics*, Undergraduate Lecture Notes in Physics,
DOI 10.1007/978-3-319-20627-1_3

thermal equilibrium at temperature T was considered, under the assumption that the inner walls of the cavity could emit radiation at all possible frequencies. By using a general thermodynamic argument, Kirchhoff demonstrated that the spectrum of the "blackbody" radiation from such a cavity depends only on T, and is independent of the nature of the material or the shape of the cavity. Defining $\rho(\nu, T)d\nu$ as the density of electromagnetic energy inside the cavity in the frequency interval between ν and $\nu + d\nu$, a very strong inconsistency between predictions based on Maxwell equations and experimental observations of $\rho(\nu, T)$ was discovered. By making the revolutionary assumption that the energy of the dipole oscillating at frequency ν can only take discrete values that are multiples of the fundamental quantum $h\nu$, Planck was able to provide the solution to this problem in 1901, obtaining the following expression:

$$\rho(\nu, T) = \frac{8\pi h\nu^3}{c^3} \frac{1}{\exp(h\nu/k_B T) - 1}. \tag{3.1}$$

Besides the Boltzmann constant of $k_B = 1.38 \times 10^{-23}$ J $\cdot$ K^{-1}, (3.1) contains the new universal constant of h, called Planck constant, having value 6.55×10^{-34} J $\cdot$ s.

By making the derivative of (3.1), one finds that the frequency ν_m corresponding to the maximum value of $\rho(\nu, T)$ is proportional to T, following the approximate relation: $\nu_m \approx 5k_B T/\text{h}$.

Introducing the peak wavelength $\lambda_m = c/\nu_m$, expressing λ_m in meters e T in Kelvin, the following relation, usually called Wien's law, is obtained:

$$\lambda_m T = 0.288 \times 10^{-2}\ \text{m} \cdot \text{K}. \tag{3.2}$$

Figure 3.1 shows the blackbody emission spectra of the Sun (5500 K) and a tungsten-filament lamp (1500 K). In the first case the peak of the emission is in the visible range, whereas in the second case a large portion of the emitted radiation is in the infrared, indicating that the incandescent lamp is a low-efficiency source of visible light.

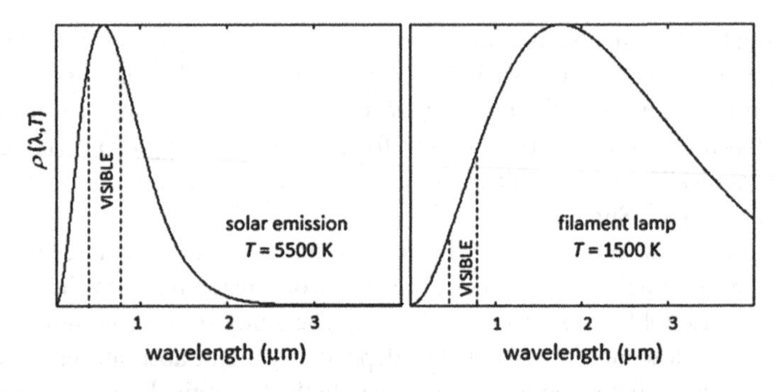

Fig. 3.1 Blackbody emission spectra, as calculated for the Sun ($T = 5500$ K) and for a tungsten-filament lamp ($T = 1500$ K). The two curves are normalized to the same maximum value

The electric-discharge lamp, also known as the fluorescent lamp, uses electron collisions to generate the atomic excitation for light emission. Notwithstanding the fact that the working temperature can be much lower, this approach provides a higher conversion efficiency of electrical power to visible light power than incandescent sources.

3.2 Origins of the Laser

Whereas traditional light sources consist of a collection of independently emitting atoms, the laser is a source in which individual emitters have a correlated phase and also well defined emission direction.

In an important paper published in 1916 Einstein considered a blackbody cavity in thermal equilibrium containing a set of two-level atoms, and showed that the balance between absorption and emission processes becomes compatible with the blackbody law only if one assumes the existence of a previously unknown process, called stimulated emission. At that time, it was not noted that this new process can result in optical amplification. It took a further four decades before proposals aimed at using stimulated emission to amplify optical beams were put forward.

The first amplifier based on stimulated emission, called maser (acronym of "Microwave Amplification by Stimulated Emission of Radiation"), used ammonia molecules in a high magnetic field, and had emission in the microwave frequency range, at 23.87 GHz. The experiment was performed at the Columbia University, New York, by Gordon, Zeiger and Townes in 1954. They sent a beam of excited ammonia molecules into a resonant cavity, obtaining a self-sustaining emission. A maser amplifier had an important role in the first demonstration of long-distance communications using the reflection of a microwave signal from an orbiting satellite in space. In this experiment, performed by a Bell Labs team led by Pierce in 1960, a maser was used as a very sensitive low-noise device to amplify the faint signals coming from the satellite.

The fundamental step for the birth of Quantum Electronics (this was the initial name given to this new field) was the invention in 1960 of the first oscillator emitting visible light, the ruby laser, created by Maiman, a researcher working at the Hughes Research Laboratories in California. Such a realization had been prepared by the theoretical studies of Townes in the United States, and Basov and Prokhorov in Russia, studies that were awarded the Nobel Prize in 1964. The acronym "laser" comes from maser, with the substitution of "light" for "microwave". The laser is indeed an oscillator, even if the acronym laser (Light Amplification by Stimulated Emission of Radiation) seems to describe an amplifier.

Starting from 1960, Photonics (this is the name now more frequently used instead of Quantum Electronics) had a very rapid development. In a few years, laser action was demonstrated in a wide variety of solid, liquid, and gaseous materials. A turning point was the invention in 1962 of the semiconductor laser; a very small device having high efficiency and long operating lifetime, which can be fabricated using the same technology developed for integrated electronic circuits and devices. Normally every type of laser emits one (or more) well-defined wavelengths that are characteristic of the material used for amplification. A very important property of semiconductor lasers is that they can be designed to work at a pre-selected wavelength, by tuning the chemical composition of the active layer.

3.3 Properties of Oscillators

An oscillator is a device that generates a sinusoidal signal with a stable amplitude and fixed frequency. The capability of producing oscillating electric fields at higher and higher frequencies has played a basic role in the development of telecommunication systems, of computers, and, in general, of all electronic instrumentation. The laser is a device that extends the working principle of the electronic oscillator to optical frequencies. The analogy with the electronic device is conceptually important, because it explains at first glance the different emission properties of a laser with respect to traditional light sources. At the same time, the analogy immediately suggests that lasers can play a very important role in the information and communication technology.

In this section the general properties of oscillators are discussed, showing that an oscillating output signal can be generated by a scheme involving the combination of amplification with positive feedback. The discussion is developed generally, without specifying the exact nature of the signal. The description of the optical amplifier and the corresponding oscillator is deferred to the following sections.

In the block scheme presented in Fig. 3.2, the input and output signals S_i e S_u may represent a displacement in the case of the mechanical oscillator, a voltage for the electronic oscillator or an electric field for a microwave or optical oscillator. The input signal S_i goes through an amplifier with gain G. The fraction β of the amplifier output is fed back and added to S_i. Therefore, in the stationary regime, the output signal S_u is given by:

$$S_u = GS_i + \beta G^2 S_i + \beta^2 G^3 S_i + \cdots \tag{3.3}$$

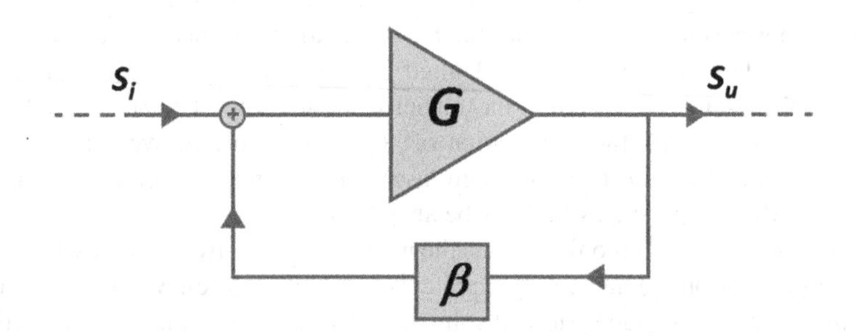

Fig. 3.2 General scheme of the oscillator: amplifier with positive feedback

By assuming that $|\beta G| < 1$, the geometric series represented by (3.3) is convergent. One obtains:

$$S_u = S_i \frac{G}{1 - \beta G}. \tag{3.4}$$

The quantity βG is called loop gain. The transfer function of the amplifier with feedback, defined as the ratio between output and input, is given by:

$$T = \frac{S_u}{S_i} = \frac{G}{1 - \beta G}. \tag{3.5}$$

The signal S_i is taken as a sinusoidal function of time with angular frequency ω. Generally speaking, both G and β are complex functions of ω. In order to make the treatment more explicit, assume that β is independent of ω and that the frequency dependence of G has a resonance at ω_o:

$$G(\omega) = \frac{G_o}{1 + i(\omega - \omega_o)\tau} \tag{3.6}$$

The power amplification $|G(\omega)|^2$ is:

$$|G(\omega)|^2 = \frac{|G_o|^2}{1 + (\omega - \omega_o)^2\tau^2} \tag{3.7}$$

The function (3.7) is a Lorentzian (from Lorentz), and represents the effective behavior of absorption (or gain) as a function of the angular frequency in most practical cases. The quantity τ^{-1} is the half-width at half-height of the Lorentzian function.

Through substitution of (3.6) inside (3.5), one finds the transfer function of the system with feedback:

$$T(\omega) = \frac{G_o}{1 + i(\omega - \omega_o)\tau - \beta G_o} = \frac{G_{fo}}{1 + i(\omega - \omega_o)\tau_f}, \tag{3.8}$$

where $G_{fo} = G_o/(1 - \beta G_o)$, and $\tau_f = \tau/(1 - \beta G_o)$. It is evident that positive feedback increases the peak gain, and, at the same time, decreases the frequency band, such that the product between frequency band and peak gain is left unchanged. Figure 3.3 shows the behavior of the power transfer function, $|T(\omega)|^2$, for increasing values of the loop gain, within the interval $0 < \beta G_o < 1$.

For βG_o tending to 1, G_{fo} tends to infinity and the bandwidth $\Gamma_f = 2\tau_f^{-1}$ tends to 0. The condition $\beta G_o = 1$ is the threshold condition for the operation of the oscillator: in fact, if G_{fo} becomes infinite, the system can provide a finite output signal even if the input is infinitesimal.

What kind of output signal is generated by such an oscillator? Since the bandwidth becomes infinitely narrow around ω_o, the output signal must be a sinusoidal oscillation at frequency ω_o. In the general case, in which β also depends on ω, the angular

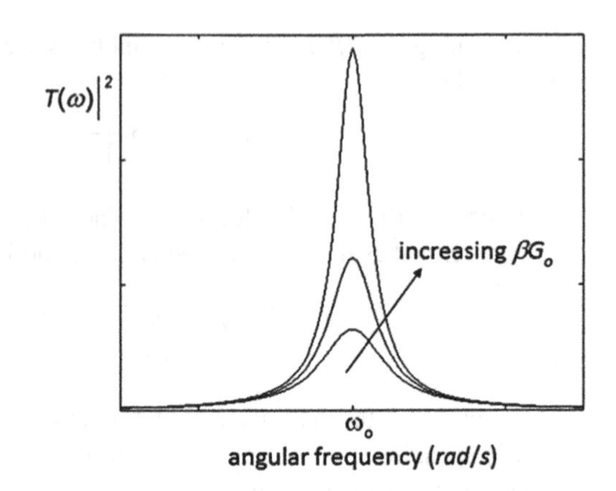

Fig. 3.3 Behavior of the frequency response of the amplifier with positive feedback for various values of the loop gain

frequency corresponding to maximum loop gain can be different from ω_o. If the loop gain is larger than 1, the oscillation is triggered by the noise always present inside a real-world loop: from amongst all the noise frequency components, the system preferentially amplifies the component with angular frequency ω_o.

The treatment developed thus far would indicate that the output signal S_u is continuously growing when the gain is above the threshold value $G_{oth} = \beta^{-1}$. This is clearly impossible, because it would violate energy conservation. Since the supply power of the amplifier has a finite value, the output power must also have a finite upper bound.

The steady-state value of S_u can be derived using a treatment that includes nonlinear effects. It is a general property of an amplifier that the gain must decrease when the amplitude of the output signal becomes large. This effect is called saturation. As shown qualitatively in Fig. 3.4, the gain G_o must be a decreasing function of S_u, going to 0 when S_u becomes large.

The working operation of an oscillator is as follows: initially the oscillator is designed to have a small-signal gain larger than β^{-1}. As long as S_u grows through

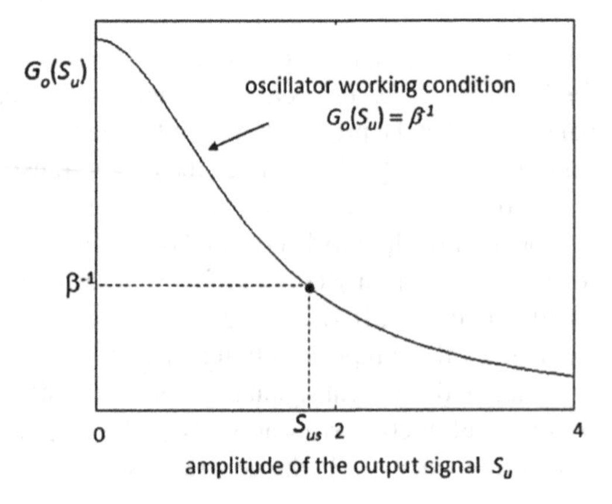

Fig. 3.4 Saturation of the amplifier: behavior of gain as a function of the output signal S_u. The working point of the oscillator is given by the condition $\beta G_o(S_{us}) = 1$

successive amplifications within the loop, the gain $G_o(S_u)$ decreases: the stationary state is reached when S_u attains the saturation value S_{us}, in correspondence of which the saturated loop gain is equal to 1. By looking at Fig. 3.4 it is evident that, if $S_u < S_{us}$ the loop gain is >1 so that S_u is growing toward the stationary value, whereas, if $S_u > S_{us}$, the loop gain becomes <1 so that S_u is now decreasing toward the stationary value. The conclusion is that the stable point of the oscillator corresponds to the situation in which the saturated loop gain is equal to 1.

3.4 Emission and Absorption of Light

The general discussion of the preceding section shows that the essential ingredient for building an optical oscillator is the optical amplifier. In this section an elementary description of the interactions between atoms and light is presented, as a preliminary step toward the treatment of the optical amplifier.

An atom is made of a nucleus with a positive electric charge at the center and a surrounding electron cloud of negative electric charge. Viewing the atom as a microscopic planetary system, its internal energy can be defined as the sum of the potential energy associated to the spatial distribution of electric charges plus the kinetic energy of the electrons that are undergoing a rotational motion around the nucleus. The configuration of the atom can be determined, at least in principle, by solving the Schrödinger equation for the system. Quantum mechanics demonstrates that the internal energy of the atom can take only discrete values, here called E_0, E_1, $E_2 \ldots$ in increasing order, where E_0 is the energy of the equilibrium configuration (ground state), and E_j, with $j > 0$, is the energy of the j-th excited state. In an intuitive model the excited state corresponds to one (or more) electrons having moved from the equilibrium orbit to an outer, higher energy, orbit. The atom may make a transition from its initial state to an upper excited state by absorbing a photon, or it may decay from an excited state to a lower-energy state by emitting a photon, but all transitions must satisfy the energy conservation relation:

$$\nu_{ik} = \frac{E_i - E_k}{h}, \tag{3.9}$$

where $E_i > E_k$. The atom in the state k can move to the state i by absorbing a photon, only if it is illuminated by light having frequency ν_{ik}. Analogously the atom decaying from state i to state k emits a quantum of light at frequency ν_{ik}. Each type of atom has different energy levels, and so it has a different discrete set of characteristic frequencies ν_{ik}. It should be mentioned that not all atomic transitions are radiative, i.e., are compatible with the absorption or emission of light. There are quantum mechanical selection rules (not discussed here) that determine which transitions are radiative.

It is useful to have in mind the order of magnitude of the energy difference for atomic transitions associated to visible light. Considering, as an example, light

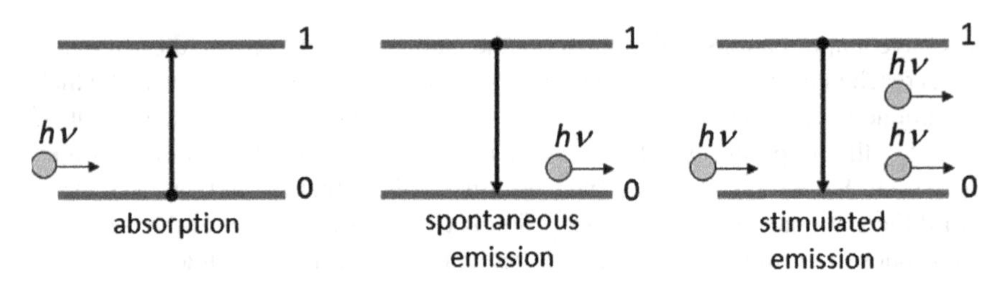

Fig. 3.5 Absorption and emission processes between levels 0 and 1

with a wavelength of $\lambda = 0.5\,\mu\text{m}$, and frequency $\nu = c/\lambda = 6 \times 10^{14}\,\text{Hz}$, the corresponding energy difference, calculated from (3.9), is 2.5 eV.

Consider a box containing N_t identical atoms per unit volume, illuminated by a monochromatic wave with intensity I and frequency ν close to ν_{10}. If the intensity is divided by the photon energy, it gives the photon flux $\Phi = I/(h\nu)$, which is the number of photons crossing the unit area in the unit time.

A schematic of the different absorption and emission processes is illustrated in Fig. 3.5. Some atoms will absorb a photon moving from the ground level 0 to level 1. The number of absorption processes per unit time and unit volume must be proportional to the number of available photons and to the number of atoms on level 0. Every absorption reduces by 1 the number of atoms on level 0, therefore:

$$-\frac{dN_0}{dt} = \sigma_{01}(\nu)\Phi N_0, \tag{3.10}$$

where $N_0(t)$ is the number of atoms per unit volume on level 0 at time t, called the population of level 0 at time t. The minus sign in front of the derivative inside (3.10) indicates that absorption processes decrease the population of level 0.

The proportionality constant $\sigma_{01}(\nu)$, called absorption cross-section, represents the probability of absorption of a photon by the atom, and has dimensions of an area. Since the energy of the excited atomic levels is always defined with some spread, $\sigma_{01}(\nu)$ is a function of ν that is peaked at ν_{10}, but has a non-zero broadening. It is typically described by a Lorentzian function.

The excited atom goes back to the ground state by spontaneously emitting a photon. The spontaneous emission is spatially isotropic, and has a spectral distribution identical to that of the absorption cross-section. It is important to point out that the radiation spontaneously emitted by a given atom has no memory of the direction, frequency or phase of the exciting radiation, and that it is uncorrelated with the emission of other nearby atoms.

By assuming that at $t = 0$ there are some excited atoms on level 1, it is found that the intensity of spontaneous emission decays exponentially with time. The time constant τ_{10}, which is characteristic of the considered transition, is called lifetime of the level 1. The reciprocal $A_{10} = \tau_{10}^{-1}$ represents the decay probability from 1 to 0 per unit time. Considering that every spontaneous emission reduces the number N_1 by 1, and that the number of spontaneous emission processes per unit time and unit volume is given by $A_{10}N_1$, the decrease of the population of level 1 is described by:

$$-\frac{dN_1}{dt} = A_{10}N_1 \tag{3.11}$$

In 1917 Einstein demonstrated that the presence of quasi-resonant radiation at a frequency close to ν_{10} stimulates the decay of the excited atom from state 1 to state 0.

In analogy with the absorption process, the number of stimulated emission processes per unit time and unit volume can be written as:

$$-\frac{dN_1}{dt} = \sigma_{10}(\nu)\Phi N_1, \tag{3.12}$$

where $\sigma_{10}(\nu)$ is the stimulated emission cross-section, which, as shown in the next section, coincides with the absorption cross-section.

If the atoms are at an absolute temperature of T, then the ratio between N_1 and N_0 is given by the Boltzmann factor:

$$\frac{N_1}{N_0} = \exp\left(-\frac{h\nu_{10}}{k_B T}\right). \tag{3.13}$$

As an example, for a transition in the red light region ($h\nu_{10} = 2\,\text{eV}$) at room temperature ($T = 300\,\text{K}$), it is found: $N_1/N_0 = e^{-80} \approx 10^{-35}$. This means that the probability of finding an excited atom is negligible, for a transition in the visible at room temperature.

If all the atoms are in the lower energy state (i.e., the level 0), a quasi-resonant light beam initially generates only absorption processes. If some fraction of the atoms are in level 0, and the rest are in level 1, then a competition arises between absorption and stimulated emission processes. If $N_0 > N_1$, absorption prevails. However, if $N_0 < N_1$, then the number of generated photons overcomes the number of absorptions. The condition of $N_0 < N_1$ is called the population inversion condition.

The important aspect of the stimulated emission process is that, in contrast to the case of spontaneous emission, the emitted radiation has the same direction, polarization, frequency and phase of the incident radiation. As such, a light beam having a frequency close to ν_{10} experiences amplification after passing through a volume containing a collection of atoms that are prepared in a state of population inversion. In order to build an optical amplifier, it is necessary to find a means of creating and maintaining a population inversion.

What happens if a set of two-level atoms is illuminated with a photon flux Φ at frequency ν_{10}? By combining (3.10), (3.11), and (3.12), and taking into account the conservation condition that $N_0 + N_1 = N_t$, the following dynamic equation for the population $N_1(t)$ is obtained:

$$\frac{dN_1}{dt} = \sigma_{01}(\nu)\Phi(N_t - 2N_1) - A_{10}N_1. \tag{3.14}$$

The first term on right-hand side of (3.14) describes the balance between absorption and stimulated emission processes, the second term takes into account spontaneous emission. Assuming that the photon flux is suddenly activated at $t = 0$ and remains constant for $t > 0$, and setting the initial condition of $N_1(0) = 0$, the integration of (3.14) gives:

$$N_1(t) = \frac{\sigma_{01}\Phi N_t}{A_{10} + 2\sigma_{01}\Phi}\{1 - \exp[-(A_{10} + 2\sigma_{01}\Phi)t]\} \tag{3.15}$$

The behavior of $N_1(t)$ is shown in Fig. 3.6. The stabilization to the steady-state response is governed by the time constant: $\tau = (A_{10} + 2\sigma_{01}\Phi)^{-1}$. The steady-state values of $N_1(t)$ and $N_0(t)$, derived for $t \to \infty$, are:

$$N_{1s} = \frac{N_t}{2 + A_{10}/(\sigma_{01}\Phi)}, \quad N_{0s} = N_t\frac{1 + A_{10}/(\sigma_{01}\Phi)}{2 + A_{10}/(\sigma_{01}\Phi)}. \tag{3.16}$$

In presence of the flux Φ the set of atoms becomes less absorbing, because the population of the ground state is reduced, and simultaneously there are stimulated emission processes partially compensating absorption. In any case, N_{1s} is always less than N_{0s}, but both stationary values tend to $N_t/2$ as Φ tends to infinity. Consequently, when $\Phi \to \infty$, the material becomes completely transparent, because the number

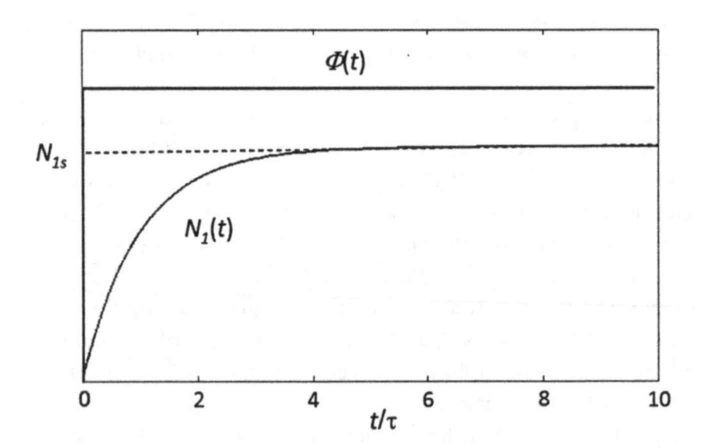

Fig. 3.6 Behavior of $N_1(t)$ in in presence of a step-function photon flux

of stimulated emission processes compensates for the absorption processes. The first of relations (3.16) can be written as:

$$N_{1s} = \frac{N_t}{2} \frac{1}{1 + \Phi_s/\Phi} \tag{3.17}$$

where $\Phi_s = A_{10}/(2\sigma_{10})$ is the saturation flux for the considered transition.

3.4.1 Einstein Treatment

In his 1917 article, Einstein not only introduced the new concept of stimulated emission, but also derived in a simple way the relations connecting emission and absorption processes. In this section the main steps of the treatment are reproduced.

Einstein considered a set of atoms inside a blackbody cavity at temperature T, observing that, at thermal equilibrium, the number of atoms per unit time making the transition from level 0 to level 1 must coincide with the number of atoms making the opposite transition. The radiation thermally generated inside the cavity is not at all monochromatic, but it presents a frequency spread comparable or even larger than the bandwidth $\Delta\nu_{10}$ of the absorption cross-section of the considered atomic transition. In such cases, in order to express the number of absorption processes, one should use, instead of (3.10) that refers to a monochromatic photon flux, the following integral:

$$\frac{dN_0}{dt} = -N_0 \int \sigma_{01}(\nu')\phi(\nu')d\nu', \tag{3.18}$$

where $\phi(\nu')$ is now the incident flux per unit frequency. The total incident flux is given by the integral:

$$\Phi = \int \phi(\nu')d\nu' \tag{3.19}$$

Of course, (3.18) reduces to (3.10) if the incident flux is monochromatic: $\phi(\nu') = \Phi\delta(\nu' - \nu)$, where $\delta(\nu' - \nu)$ is Dirac delta function. In contrast, if the bandwidth of $\phi(\nu')$ is much larger than $\Delta\nu_{10}$, then (3.18) can be approximated as:

$$\frac{dN_0}{dt} = -c^{-1} B_{01}\phi(\nu_{10})N_0, \tag{3.20}$$

where

$$B_{01} = c \int \sigma_{01}(\nu')d\nu' \tag{3.21}$$

is the absorption coefficient (expressed in units: m^3s^{-2}).

Balancing spontaneous emission and absorption processes, and introducing the energy density $\rho(\nu) = (h\nu)\phi(\nu)/c$ instead of the photon flux, one finds:

$$h\nu_{10}A_{10}N_1 = \rho(\nu_{10})B_{01}N_0 \tag{3.22}$$

Taking into account that N_0/N_1 is given by (3.13), (3.22) gives a dependence of $\rho(\nu)$ on temperature that, rather surprisingly, is different from the expression (3.1). At this point, Einstein made the hypothesis that a third process must be occurring, which he called stimulated emission. By expressing the number of stimulated emission processes per unit time with a relation similar to that used for absorption processes, (3.12) transforms into:

$$\frac{dN_1}{dt} = -c^{-1}B_{10}\phi(\nu)N_1 \tag{3.23}$$

where B_{10} is the stimulated emission coefficient:

$$B_{10} = c\int \sigma_{10}(\nu')d\nu' \tag{3.24}$$

The balance between absorption and emission, including now also stimulated emission, becomes:

$$h\nu_{10}A_{10}N_1 + \rho(\nu_{10})B_{10}N_1 = \rho(\nu_{10})B_{01}N_0. \tag{3.25}$$

From (3.25) the following expression is derived:

$$\rho(\nu_{10}) = \frac{h\nu_{10}A_{10}}{B_{01}(N_0/N_1) - B_{10}}. \tag{3.26}$$

By taking into account (3.13), and comparing (3.26) with (3.1), the following relations are obtained:

$$B_{10} = B_{01}\ , \quad \frac{A_{10}}{B_{01}} = \frac{8\pi\nu_{10}^2}{c^3} \tag{3.27}$$

Note that the equality $B_{10} = B_{01}$ implies that $\sigma_{10} = \sigma_{01}$.

Relations (3.27) show symmetry between absorption and stimulated emission processes, and demonstrate that the spontaneous emission coefficient A_{10} is proportional to B_{01}. The quantum-mechanical calculation, not discussed here, shows that B_{01} is proportional to ν_{10}. As a consequence, A_{10} is proportional to the cube of the frequency: this means that the spontaneous decay time $\tau_{10} = (A_{10})^{-1}$ is much longer for infrared transitions with respect to visible transitions, and becomes very short for UV or soft X-ray transitions. As will be seen, very short decay times of the excited level make it much more difficult to create and maintain a population inversion for a given transition, and so it is comparatively difficult to build optical amplifiers and lasers in the UV or soft X-ray region.

3.5 Optical Amplification

In this section it is described how to obtain optical amplification. It is clear from the previous sections that it is impossible to obtain a steady-state population inversion on the transition 0–1 using an approach dealing only with levels 0 and 1. The excitation process must involve a third level.

Three-level system. The discussion starts by considering a three-level system, like the one shown in Fig. 3.7. The scheme is the same as that used in the first laser; the ruby laser. The aim is to obtain a population inversion between the ground level 0 and the excited level 1. In the jargon of Photonics the process producing population inversion is called pumping process, and the amplifying medium is called active medium. The idea is to pump with light resonant with the transition between level 0 and level 2. In presence of a photon flux $\Phi_p(\nu)$ at ν_{02}, absorption processes exciting atoms to level 2 occur. The spontaneous decay from level 2 to level 1 populates level 1. Intuitively, one could predict that population inversion can be obtained if the spontaneous decay process from 2 to 1 is quicker than the process from 1 to 0. In a regime of dynamic equilibrium, this does indeed create a "bottleneck" that favors the growth of N_1.

The quantitative treatment of the three-level scheme is based on the so-called rate equations. These describe the population dynamics by taking into account all possible emission and absorption processes among the three levels:

$$\frac{dN_1}{dt} = A_{21}N_2 - A_{10}N_1 \tag{3.28}$$

$$\frac{dN_2}{dt} = W_{02p}(N_0 - N_2) - (A_{20} + A_{21})N_2, \tag{3.29}$$

where A_{10}, A_{20} and A_{21} are spontaneous decay probabilities per unit time, and W_{02p} is the probability per unit time of a transition from level 0 to level 2 under the action of the pump. Using (3.18), one finds: $W_{02p} = \int \sigma_{02}(\nu')\phi_p(\nu')d\nu'$.

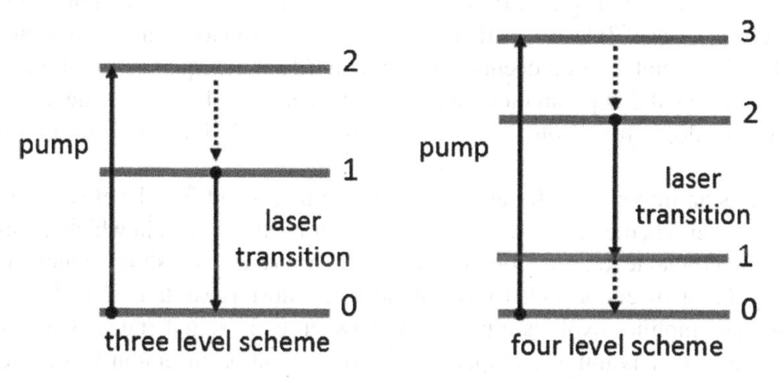

Fig. 3.7 Three- and four-level schemes. The *vertical direction* indicates increasing energy of the atomic levels

Since the sum of the three populations coincides with the total number of atoms per unit volume N_t:

$$N_0 + N_1 + N_2 = N_t, \tag{3.30}$$

N_0 can be eliminated from (3.29), so that a system of two equations with two unknown functions is obtained. If W_{02p} is constant, then the system can be easily solved. Here the discussion is limited to the steady-state solutions, obtained by putting all the time-derivatives equal to 0.

The steady-state population inversion is given by:

$$N_s = N_{1s} - N_{0s} = N_t \frac{1 - \frac{A_{10}}{A_{21}} \frac{W_{02p}+A_{20}+A_{21}}{W_{02p}}}{1 + \frac{A_{10}}{A_{21}} \frac{2W_{02p}+A_{20}+A_{21}}{W_{02p}}}. \tag{3.31}$$

In order to obtain amplification, N_s should be positive, which means that the sign of the fraction appearing at right-hand side of (3.31) must be positive. For very large values of the pump parameter, $W_{02p} \gg A_{20} + A_{21}$, it is a necessary condition that $A_{21} > A_{10}$, in agreement with the intuitive considerations mentioned above. It is interesting to note that, if $A_{21} \gg A_{10}$ and simultaneously $W_{02p} \gg A_{20} + A_{21}$, then (3.31) reduces to $N_s = N_t$. This means that all the atoms have been excited to level 1.

The condition $N_s = 0$ gives the threshold value of the pump parameter:

$$W_{02pth} = A_{10} \frac{A_{20} + A_{21}}{A_{21} - A_{10}}. \tag{3.32}$$

To avoid excessive values of W_{02pth}, systems in which A_{21} is much larger than A_{10} are usually chosen. The first laser (i.e., the ruby laser) has the simple three-level scheme, with A_{21} many orders of magnitude larger than A_{10}.

Four-level system. Optical amplification can be also achieved by using a four-level scheme, illustrated in Fig. 3.7. In this scheme, the pump excites the atom from the ground level to level 3; level 2 is then populated by the spontaneous decay from 3 to 2, and optical amplification occurs between 2 and 1, at a frequency ν_{21}. Intuitively it can be expected that population inversion between 2 and 1 can be achieved, if the spontaneous decay time from 3 to 2 is short, from 2 to 1 is long, and from 1 to 0 is short.

There is an important difference between the three- and four-level schemes. In the three-level scheme, the amplifier exploits the transition 1 to 0, in which the lower level is the ground level. As a consequence, in order to reach threshold, which means zero population inversion, at least half of the atoms must be excited. In the four-level scheme the amplifier exploits a transition between two excited states. This means that, since level 1 is initially empty, a positive population inversion is technically achieved when even only one atom is brought to level 2. As a result, the four-level scheme requires much lower pump power with respect to the three-level scheme for lasing, and so most practical lasers operate using a four-level scheme.

The dynamics of the four-level system is described by the following rate equations:

$$\frac{dN_3}{dt} = W_{03p}(N_0 - N_3) - (A_{30} + A_{31} + A_{32})N_3 \tag{3.33}$$

$$\frac{dN_2}{dt} = A_{32}N_3 - (A_{20} + A_{21})N_2 \tag{3.34}$$

$$\frac{dN_1}{dt} = A_{31}N_3 + A_{21}N_2 - A_{10}N_1, \tag{3.35}$$

where $W_{03p} = \int \sigma_{03}(\nu')\phi_p(\nu')d\nu'$.

The system (3.33)–(3.35) is completed by:

$$N_0 + N_1 + N_2 + N_3 = N_t. \tag{3.36}$$

Again, the discussion is limited to the steady-state solutions, by putting all time derivatives equal to 0. The steady-state population inversion is found to be:

$$N_s = N_{2s} - N_{1s} = N_t \frac{(1 - C_1)W_{03p}}{C_2 + C_3 W_{03p}}, \tag{3.37}$$

where

$$C_1 = \frac{A_{31}A_{21} + A_{31}A_{20} + A_{32}A_{21}}{A_{10}A_{32}} \tag{3.38}$$

$$C_2 = \frac{(A_{21} + A_{20})(A_{30} + A_{31} + A_{32})}{A_{32}} \tag{3.39}$$

$$C_3 = \frac{A_{31}A_{21} + A_{31}A_{20} + A_{32}A_{10} + 2A_{10}A_{21} + 2A_{10}A_{20}}{A_{10}A_{32}}. \tag{3.40}$$

As expected from intuitive considerations, the threshold pump parameter W_{03pth} is equal to 0. The steady-state population inversion N_s becomes positive when $C_1 < 1$. In many practical cases $A_{20} \ll A_{21}$ and $A_{31} \ll A_{32}$, so that the condition $N_s > 0$ simply reduces to: $A_{21} < A_{10}$.

Since the rate equations constitute a set of linear first-order differential equations, one can determine analytically the complete time-dependent solutions, once the initial conditions are given. The important feature of these solutions is that they give the response time of the atomic system τ_r. The atomic system can follow variations of the pump power only if they happen on a time-scale that is not shorter than τ_r.

Optical amplifier. Consider the propagation of a monochromatic light beam, at frequency ν close to ν_{21}, inside an amplifying medium made of four-level atoms.

Assuming that the beam propagates along the z axis and that the amplifier extends from $z = 0$ to $z = L$, the infinitesimal variation of the photon flux Φ over dz at z is given by the following balance between stimulated emission and absorption:

$$d\Phi = \sigma(\nu)(N_2 - N_1)\Phi dz. \tag{3.41}$$

For simplicity of notation σ appears instead of σ_{12}.

In the small-signal regime the photon flux can be considered sufficiently weak so as not to appreciably modify the populations of atomic levels. If $N_2 - N_1$ is independent of z, then (3.41) becomes a linear equation, and its integration gives:

$$\Phi(z) = \Phi_o\, e^{g(\nu)z}, \tag{3.42}$$

where Φ_o is the input flux at $z = 0$, and $g(\nu) = \sigma(\nu)(N_2 - N_1)$ is the gain per unit length. The total gain, calculated as the ratio output to input, is:

$$G = e^{g(\nu)L} \tag{3.43}$$

Note that (3.41) does not contain the contribution of spontaneous emission, because spontaneous emission is spread over all directions and all frequencies within the transition bandwidth. As such, it can be assumed that very few photons are spontaneously emitted along z at frequency ν. Spontaneous emission is an intrinsic noise source for the optical amplifier. Concerning noise, the difference between the electronic amplifier and the optical amplifier is that for an electronic amplifier $h\nu$ is smaller than $k_B T$ (i.e. the noise is essentially thermal), while for the optical amplifier $h\nu$ is much larger than $k_B T$, so that thermal noise is negligible.

3.6 Scheme and Characteristics of the Laser

As discussed in Sect. 3.3, an oscillator is an amplifier inserted in a positive feedback scheme. Therefore, in order to make a laser (i.e., an optical oscillator), one should create a feedback loop in which some part of the amplifier output is returned to the input. Two typical schemes are shown in Fig. 3.8. In the ring cavity the optical beam follows a closed loop, propagating either clockwise or counter-clockwise. In the two-mirror scheme, the two flat or curved mirrors are set up facing each other, so that an optical wave can bounce back and forth between the two mirrors. Such a cavity is known as Fabry-Perot cavity, because it has the same structure of the Fabry-Perot interferometer treated in Sect. 2.4.4. Compared to the ring cavity, the Fabry-Perot cavity is simpler, more compact, and has the advantage that, in a single round-trip, the laser beam crosses twice the amplifier. Usually the amplifier has a cylindrical geometry and the laser beam propagates along the axis of the cylinder. One of the mirrors is partially transmitting, so that a fraction of the optical power traveling inside the cavity is coupled out as the laser emission.

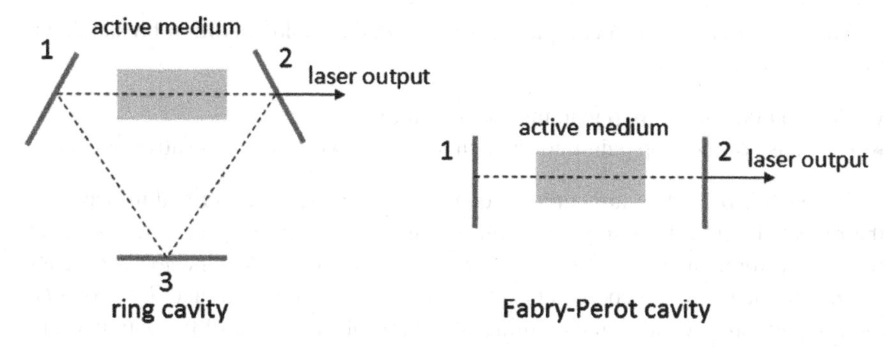

Fig. 3.8 Laser configurations. Mirrors 1 and 3 are totally reflecting, the laser output comes from mirror 2 that is partially transmitting

Because of the finite transverse dimensions of the active medium and mirrors, a light beam that travels even slightly off-axis will leave the cavity region after a given number of passes between the two mirrors. Therefore, in order to be amplified by many passes, the laser beam must travel very closely along the cavity axis. This results in a highly directional laser output.

The laser threshold condition corresponds to the situation in which the loop gain is equal to 1. Introducing an idealized signal, $E(t)$, into the cavity, and then determining how the signal changes after a single one-loop pass, allows the loop gain to be deduced.

The calculation is performed by considering a Fabry-Perot cavity with a distance L between mirrors, and an electric-field reflectivity of ρ_1 and ρ_2 for mirror-1 and mirror-2, respectively. The amplifier is made of a material with refractive index n, has a length l, and provides a gain G for the optical intensity (or photon flux). In order to fully understand the behavior of the laser cavity it is necessary to consider the light beam as a wave, possessing amplitude and phase, instead of a stream of photons. Taking into account that the intensity is proportional to the modulus-square of the field, and that a wave of angular frequency ω, traveling for a distance z in a medium with index of refraction n, undergoes a phase shift of $(\omega/c)nz$, then the effect of the amplifier is that of multiplying the electric field of the input wave by a complex factor $\sqrt{G}\exp[i(\omega/c)nl]$. By considering also the free propagation in air over the distance $L - l$ and the reflections on the two mirrors, the electric field $E'(t)$ of the wave that has completed a round trip is given by:

$$E'(t) = E(t)\rho_1\rho_2 G \exp[i(2\omega/c)(L - l + nl)]. \tag{3.44}$$

The threshold condition, that is the unitary loop-gain condition, is obtained by putting $E' = E$:

$$\rho_1\rho_2 G\exp[i(2\omega/c)(L - l + nl)] = 1. \tag{3.45}$$

Since the loop gain is a complex quantity, the threshold condition breaks down into two conditions:

- The modulus of the loop gain must be equal to 1.
- The phase of the loop gain must be equal to $2q\pi$, where q is a positive integer.

In general, ρ_1 and ρ_2 are complex quantities, because, as described in Sect. 2.2, the reflected field differs in both amplitude and phase with respect to the incident field. Therefore, one may have: $\rho_1 = \sqrt{R_1}\exp(i\phi_1)$, $\rho_2 = \sqrt{R_2}\exp(i\phi_2)$, where R_1 are R_2 are the reflection coefficients for the optical power. Given that $G = \exp(gl)$, the unity modulus condition determines the threshold value g_{th} of the gain per unit length inside the amplifier:

$$\sqrt{R_1 R_2}\exp(g_{th}l) = 1. \tag{3.46}$$

The phase condition defines a discrete set of oscillation frequencies, ν_q, usually called longitudinal modes. Taking $L' = L - l + nl$ as the optical path between mirror 1 and mirror 2, the following expression is derived from (3.45):

$$\nu_q = \frac{c}{2L'}(q - \frac{\phi_1 + \phi_2}{2\pi}). \tag{3.47}$$

This coincides with the relation (2.119), derived in Sect. 2.4.4 for the resonance frequencies of the Fabry-Perot interferometer.

The frequency difference between two adjacent modes, $\Delta\nu = \nu_{q+1} - \nu_q$, is:

$$\Delta\nu = \frac{c}{2L'}. \tag{3.48}$$

As it would be expected, this relation reproduces (2.120). Note that $2L'$ is the optical path corresponding to one round trip inside the laser cavity, therefore $2L'/c$ is the time taken by light to make one round trip. Consider a numerical example: if $L' = 7.5\,\text{cm}$, $2L'/c = 0.5\,\text{ns}$, and $\Delta\nu = 2\,\text{GHz}$.

Figure 3.9 qualitatively shows $g(\nu)$, which is proportional to the stimulated-emission cross-section σ_{21}, and so is a function centered at ν_{21}, with a half-height width $\Delta\nu_{21}$. The figure refers to a laser operating above threshold, because the peak value of $g(\nu)$ is larger than g_{th}. The laser has two longitudinal modes, corresponding to two auto-frequencies, running in the cavity. In general, the laser will operate only if at least one of the auto-frequencies ν_q belongs to the frequency interval over which the above-threshold condition is satisfied. This will certainly happen if $\Delta\nu$ is smaller than $\Delta\nu_{21}$.

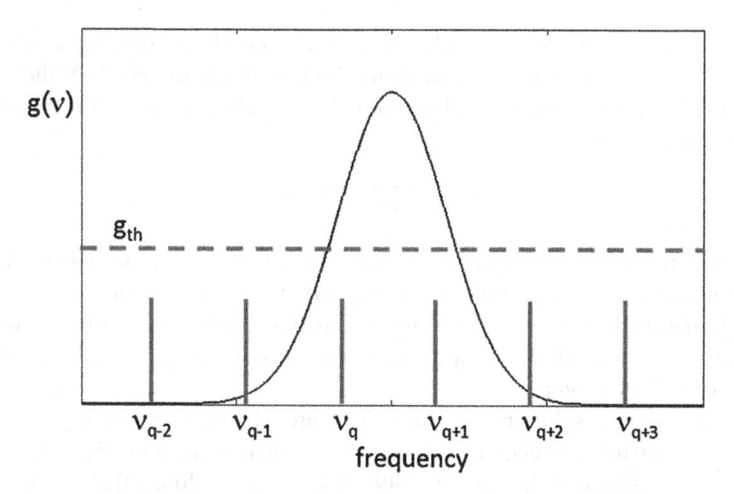

Fig. 3.9 Gain per unit length as a function of frequency. A set of auto-frequencies is also shown

3.7 Rate Equations

The general discussion of Sect. 3.3 indicates that the steady-state behavior of the oscillator above threshold is determined by the gain-saturation effects. In this section a simple nonlinear laser model is discussed, in which saturation effects are included.

The model considers an amplifier working on a four-level scheme, inserted into a Fabry-Perot cavity and operating on a single longitudinal mode. The treatment is greatly simplified by assuming that the dynamic behavior of the laser is determined by only two time-dependent quantities, the total number of photons inside the cavity, $Q(t)$, and the population of the upper laser level $N_2(t)$. The implicit assumption is that level 3 has a very fast spontaneous decay toward level 2, and level 1 is rapidly decaying toward level 0, so that N_1 is close to 0 and N_2 can be taken to coincide with the population inversion. The nonlinear equations expressing the balance between gain and losses are the following:

$$\frac{dQ}{dt} = VBQN_2 - \frac{Q}{\tau_c} \tag{3.49}$$

$$\frac{dN_2}{dt} = W - BQN_2 - \frac{N_2}{\tau}, \tag{3.50}$$

where B is proportional to the stimulated emission coefficient, V is the volume occupied by the laser mode inside the active medium and W is the number of atoms brought to level 2 by the pump per unit time and volume. W is proportional to the pump power P_p. The time constant τ is the spontaneous decay time of level 2, and τ_c represents the decay time of the photon number inside the cavity in absence of pump. In general, τ_c, sometimes called photon lifetime of the laser cavity, depends on all cavity losses, including, besides the useful loss of the semi-transparent output mirror, also losses due to diffraction, to misalignments and to defects of the optical components. If the only important loss is due to the output mirror having reflectance R, then τ_c is expressed as:

$$\tau_c = \frac{2L'}{c(1-R)}. \tag{3.51}$$

The two rate equations have a very simple meaning. Equation (3.49) indicates that the total number of photons inside the cavity increases because of stimulated emission, but decreases because of losses. Equation (3.50) says that the number of atoms on the upper level N_2 increases because of the pumping process, but decreases because of the emission processes.

The steady-state solutions, Q_s and N_{2s}, are obtained by putting the time-derivatives equal to 0. In general, nonlinear systems may admit more than one steady-state solution. Indeed, in the case of (3.49) and (3.50), it is found that two solutions exist:

$$\begin{aligned} Q_s &= 0 \\ N_{2s} &= W\tau, \end{aligned} \tag{3.52}$$

and

$$\begin{aligned} Q_s &= V\tau_c(W - W_{th}) \\ N_{2s} &= (VB\tau_c)^{-1}, \end{aligned} \tag{3.53}$$

where

$$W_{th} = (VB\tau\tau_c)^{-1}. \tag{3.54}$$

By performing a stability analysis on the pump parameter W, it is possible to determine which of the solutions is stable and therefore has a real physical basis. The solution (3.52) is stable, provided that $W \leq W_{th}$. In this situation, the value of the pump parameter is too small to compensate for cavity losses. The loop gain is then smaller than 1 and so the laser is below threshold. Note that below threshold the laser cavity contains the photons due to spontaneous emission processes, and even if Q_s is small, it will not be exactly zero. The mathematical result of $Q_s = 0$ is due to the fact that the contributions of spontaneous emission do not appear in the above simplified rate equations.

As the parameter W increases, there is an exchange of stability between the two solutions. The solution (3.53) is stable, provided that $W > W_{th}$, i.e., if the loop gain is larger than 1, meaning that the laser is above threshold.

The solution (3.53) also indicates that, when the laser is above threshold, the population of the upper level does not grow as W is increased, but remains clamped to the threshold value: this is consistent with the fact that in the steady-state regime the saturated loop-gain should be equal to 1. Below threshold the pump power is

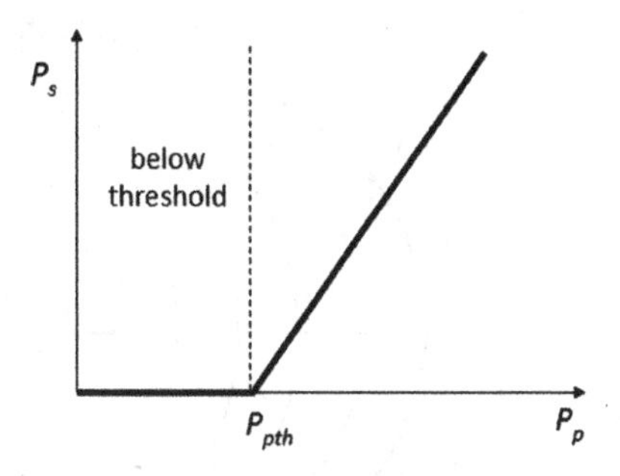

Fig. 3.10 Laser output power as a function of pump power. P_{pth} is the threshold pump power

completely used to populate the upper level of the atomic transition until the loop gain inside the cavity approaches 1. Above threshold the fraction of pump power that exceeds the threshold value is converted into laser power.

With the assumption that cavity losses are due only to the transmission through the output mirror, the laser output power is given by:

$$P_s = h\nu \frac{Q_s}{\tau_c}. \tag{3.55}$$

The behavior of P_s as a function of the pump power is presented in Fig. 3.10.

An important parameter is the laser efficiency, defined as the ratio between output optical power and pump power:

$$\eta = \frac{P_s}{P_p}. \tag{3.56}$$

For the first laser, i.e., the ruby laser, η was of the order of 10^{-3}. As it will be seen, there are lasers for which η can approach 0.5.

As a final comment to this section, it should be emphasized that the treatment has implicitly assumed that the active medium is made of a dilute set of atoms. In other words, the atoms are sufficiently far apart one from the other so that the energy level structure of the isolated atom is not modified. Most existing lasers are consistent with this assumption, but there are exceptions, the most important of which are semiconductor lasers. In any case, even if some change in the structure of the rate equations needs to be introduced, the main conclusions drawn from the discussion of (3.49) and (3.50) remains largely valid for all types of lasers.

3.8 The Laser Cavity

The two finite-size mirrors of the Fabry-Perot laser constitute an open cavity that is intrinsically lossy, because of diffraction. A general property of optical cavities is that there is a discrete number of possible field configurations, known as cavity modes. In this section the discussion is limited to Fabry-Perot cavities made with

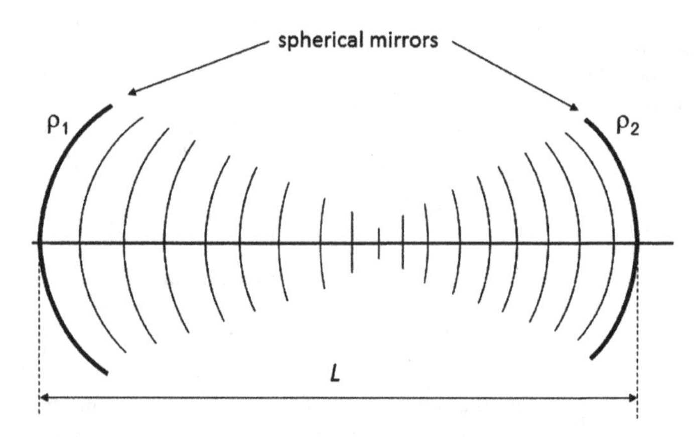

Fig. 3.11 Laser cavity made by two spherical mirrors at distance L

spherical mirrors. A very remarkable result of the treatment is that, within the paraxial approximation, the cavity modes are Gaussian spherical waves.

The cavity, as shown in Fig. 3.11, is made by two spherical mirrors at distance L. The radii of curvature of the mirrors, called ρ_1 and ρ_2, are considered positive if the mirror concavity is directed inwards in the cavity. Gaussian spherical waves are possible candidates as cavity modes because their shape is invariant in free propagation and in reflection from a spherical mirror, as discussed in Sect. 1.4.2. The general condition ensuring that the wave reflected by a spherical mirror retraces back exactly the incident wave is that the radius of curvature of the wavefront of the incident wave coincides with the radius of curvature of the spherical mirror surface. If such a condition can be satisfied for both mirrors, the Gaussian spherical wave is then reflected back and forth between the two mirrors without a change in the transverse profile. By assuming that the beam waist inside the cavity is at distance z_m from mirror 1, and recalling that the z-dependence of the radius of curvature of the wavefront is given by (1.39), the two conditions are:

$$z_m + \frac{{z_R}^2}{z_m} = \rho_1 \; ; \quad L - z_m + \frac{{z_R}^2}{L - z_m} = \rho_2 \tag{3.57}$$

The two unknown quantities inside (3.57) are z_m and z_R, or z_m and w_o, if one recalls that w_o is related to z_R by (1.38). Introducing the two parameters $g_1 = 1 - L/\rho_1$ and $g_2 = 1 - L/\rho_2$, the solutions of (3.57) are expressed as follows:

$$z_m = \frac{g_2(1 - g_1)}{g_1 + g_2 - 2g_1g_2} L \; ; \quad z_R = \frac{\sqrt{g_1g_2(1 - g_1g_2)}}{g_1 + g_2 - 2g_1g_2} L. \tag{3.58}$$

Noting that, for a symmetrical cavity, $g_1 = g_2$, one finds from (3.58) that $z_m = L/2$, which means that the waist of the Gaussian spherical wave is in the center of the cavity.

The beam radii w_1 and w_2 at the two mirrors are given by:

$$w_1{}^2 = \frac{L\lambda}{\pi}\sqrt{\frac{g_2}{g_1(1-g_1g_2)}}; \quad w_2{}^2 = \frac{L\lambda}{\pi}\sqrt{\frac{g_1}{g_2(1-g_1g_2)}}. \tag{3.59}$$

Up to this point, only the spatial dependence of the mode was considered. The second step is that of determining the resonance frequencies, by using the requirement that the complete round-trip phase delay be a multiple of 2π. The calculation must take into account that the Gaussian spherical wave acquires during propagation an additional phase delay $\psi(z)$, given by (1.37), with respect to the plane wave. The resonance condition reads:

$$2\left[\frac{2\pi\nu_q}{c}L - \psi(z_m) + \psi(L - z_m)\right] + (\phi_1 + \phi_2) = 2q\pi, \tag{3.60}$$

where q is a positive integer, and ϕ_1, ϕ_2 are the phase shifts due to the reflections from the mirrors. By combining (3.58) and (3.60), it is finally found:

$$\nu_q = \left(q - \frac{\phi_1 + \phi_2}{2\pi} + \frac{\arccos\sqrt{g_1g_2}}{\pi}\right)\frac{c}{2L} \tag{3.61}$$

If compared with (3.47), (3.61) shows that the Gaussian wave approach produces a rigid shift of the resonance frequencies found with the plane wave treatment, keeping unchanged the frequency difference between adjacent resonances.

Equation (3.61) gives real resonance frequencies only if the condition

$$0 \leq g_1g_2 \leq 1 \tag{3.62}$$

is satisfied.

The condition (3.62) can be conveniently displayed in the g_1–g_2 plane, as shown in Fig. 3.12. The shaded areas in the diagram of Fig. 3.12, in which the condition (3.62) is fulfilled, correspond to stable cavities, and the clear areas, in which (3.62) is violated, correspond to unstable cavities. Unstable cavities are very lossy, therefore lasers normally use stable cavities.

It is instructive to discuss a numerical example. Considering a stable symmetric cavity with $g_1g_2 = 0.5$, the beam waist of the Gaussian wave, calculated from (3.58), is: $w_o \approx 1.1\sqrt{L\lambda/\pi}$. Therefore, the Gaussian wave emitted by such a laser cavity has a divergence angle given by: $\theta_o = \lambda/(\pi w_o) \approx 0.9\sqrt{\lambda/\pi L}$. Taking, for instance, $\lambda = 500\,\text{nm}$ and $L = 10\,\text{cm}$, it is found: $w_o \approx 0.14\,\text{mm}$ and $\theta_o \approx 1.14 \times 10^{-3}\,\text{rad}$.

It should be recalled that the full theory based on the paraxial approximation predicts the existence of generalized Gaussian modes presenting a more complicated transverse field distribution, as the Hermite-Gauss modes described in Sect. 1.4.2. The presence of the higher-order transverse modes in the laser output is considered in almost all cases a nuisance. It is a general fact that the higher-order transverse modes extend farther out in the transverse direction and have more of their energy at greater

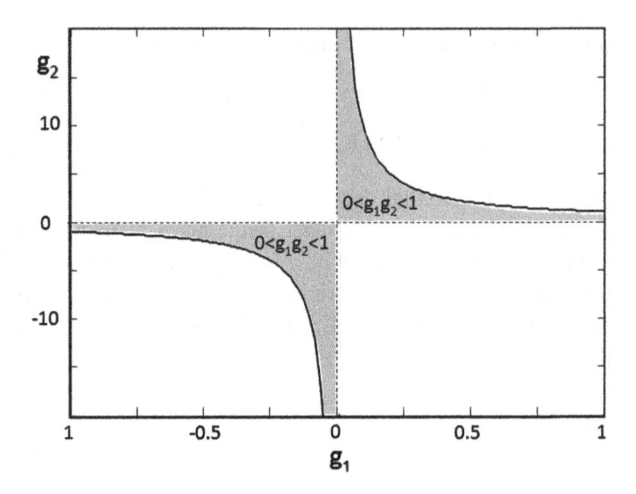

Fig. 3.12 A g_1–g_2 stability diagram for a general spherical laser cavity

distances from the axis than does the Gaussian mode. Suppose the transverse size of a cavity is defined by a limiting aperture of diameter $2a$, which can be the mirror (or the laser rod) diameter or even an iris placed inside the cavity for mode-control purpose. All the mode losses will increase as the aperture radius a is reduced toward the mode size w. This loss will be larger for the higher-order modes because of their larger transverse spread. The single-mode condition can be expressed by stating that the quantity $N_F = a^2/(\lambda L)$, known as the Fresnel number of the cavity, should have a value around 1. In the common terminology, a laser operating on a single transverse mode, i.e., emitting a Gaussian wave, is said to work on the TEM_{00} mode, where TEM stays for "transverse electromagnetic".

3.9 Solid-State and Gas Lasers

It has been shown in the previous sections that it is possible to obtain optical amplification by exciting specific energy levels of atoms. Ions, which are atoms that have lost or gained one or more electrons, can be also candidates for laser action.

Molecules have an internal energy due not only to their electronic structure, but also to their vibrational and rotational motion. As predicted by quantum mechanics, molecules present discrete vibrational and rotational energy levels, and can make transitions among these levels by absorbing or emitting electromagnetic radiation. While lasers based on electronic transitions emit in the visible or near infrared, molecular lasers typically emit in the middle infrared, because the energy jumps associated with vibrational and rotational transitions are of the order of 0.1 and 0.01 eV, respectively.

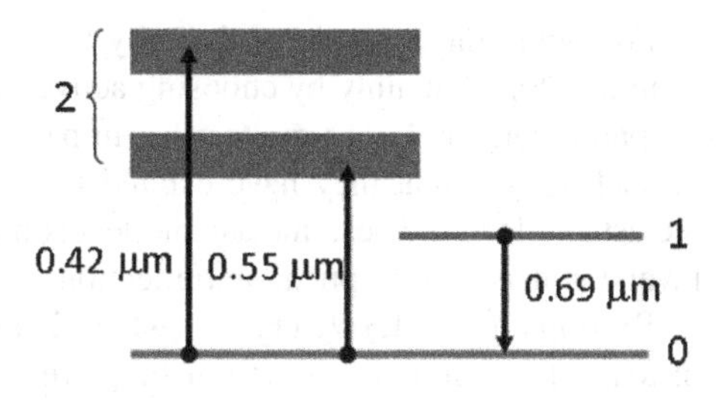

Fig. 3.13 Ruby laser: simplified scheme of the energy levels of the ion Cr^{3+}

There are also lasers utilizing transitions that are both electronic and vibrational (termed "vibronic"). These lasers are interesting because they can be tuned to emit within a rather wide band of wavelengths.

In this section, only solid-state and gas lasers are considered. The descriptions of semiconductor lasers and fiber-optic lasers are deferred to Chaps. 5 and 6, respectively.

Solid-state lasers. The term solid-state laser indicates a laser using a dielectric crystal (or a glass) doped with a small quantity of impurities as the active medium. The first laser ever made (i.e., the ruby laser in 1960) was a solid-state laser. Ruby is a piece of Al_2O_3 crystal, in which a small fraction of Al^{3+} ions are replaced by chromium ions. In the ruby crystals used as active media the percentage of Cr_2O_3 is typically 0.05 % by weight. The pure Al_2O_3 crystal, called corundum, by itself is perfectly transparent in the visible. It appears red when doped with chromium, because Cr^{3+} ions have two absorption bands in the visible range, one in the green and the other in the violet, as shown in the simplified scheme of Fig. 3.13.

The ruby laser is a three-level laser: level 0 is the ground level, level 1 is the first excited level, level 2 is represented by the two absorption bands. The pump power comes from a lamp, emitting light on a broad spectrum that includes the green and violet wavelengths required to excite the chromium ions to level 2. Level 1 is populated by spontaneous decay processes from level 2. The lifetime of level 2 is of the order of picoseconds, while that of level 1 is of the order of milliseconds. As shown in Sect. 3.5, it is possible to create population inversion between 1 and 0 under such conditions. The fact that level 2 is made by two relatively broad bands instead of a sharp energy level is very helpful, because it ensures a larger absorption of the broad-band light emitted by the lamp. The energy difference between excited and ground level is $E_1 - E_0 = 1.79\,\text{eV}$, corresponding to red light with wavelength $\lambda = 0.6943\,\mu\text{m}$.

The active medium is a cylindrical monocrystal with a length of a few centimeters and diameter of a few millimeters. In the first laser the two end faces of the ruby bar were made flat and parallel, and were coated with a thin silver layer, in order to create cavity mirrors. The pump light was provided by a helical xenon flashtube surrounding the active rod. The ruby laser has a high threshold power, a low efficiency (typically, $\eta \approx 0.1\,\%$), and exhibits large fluctuations in output power. Consequently, although it is of historical and illustrative importance, the ruby laser is no longer used in modern applications.

Following the example of the ruby laser, many different solid-state lasers have been developed, mainly by choosing active ions that belong to the transition metals (in particular, Cr, Ti), or rare earths (in particular, Nd, Er, Yb). A common feature of such ions is that they have excited levels with very long decay times, termed metastable levels. These metastable levels make it easy to build up the population inversion needed for optical amplification.

By doping the Al_2O_3 crystal (which is also known as sapphire) with titanium instead of chromium, a titanium-sapphire laser material can be produced. When pumped by green light, this laser is tunable over a very broad wavelength range, from 700 to 1000 nm, by exploiting a vibronic transition.

The most important solid-state laser is the neodymium laser, in which the host medium is often a YAG (yttrium aluminium garnet, or $Y_2Al_5O_{12}$) crystal, in which some of the Y^{3+} ions are replaced by Nd^{3+} ions. Besides YAG, other host media, such as yttrium vanadate (YVO_4), can be used.

The simplified scheme of the energy levels of the ion Nd^{3+} in YAG is shown in Fig. 3.14. The main difference with respect to the ruby laser is that the laser transition involves two excited levels, that is, one deals with a four-level scheme. The two absorption bands that play the role of level 3 are centered on the wavelengths 0.73 and 0.81 μm. These bands are coupled by a fast non-radiative decay ($A_{32} \approx 10^7\ \mathrm{s}^{-1}$) to level 2, which is the upper level of the laser transition. The spontaneous decay time from 2 to 1 is of the order of hundreds of microseconds, much longer than the spontaneous decay time from 1 to 0. The laser emission is in the infrared, at $\lambda = 1.064\ \mu\mathrm{m}$. The structure of the Nd:YAG laser is similar to that of the ruby laser. The Nd:YAG laser has a much lower threshold power and a much larger efficiency (η around 1–3 %) when compared to the ruby laser. While the ruby laser can only work as a pulsed laser, the Nd laser can operate both as a continuous wave (CW) laser with output power up to hundreds of watts, and a pulsed laser with a pulse repetition rate of 10–100 Hz. The half-height gain bandwidth is around 200 GHz, and so the laser operates simultaneously on many longitudinal modes, when it has a Fabry-Perot cavity with a length of a few centimeters.

In the early times of solid-state lasers, the pump light was provided by electric discharge lamps. The modern trend is to use, instead of lamps, another laser, the semiconductor laser. In the case of the Nd laser, the optimum pump source is the

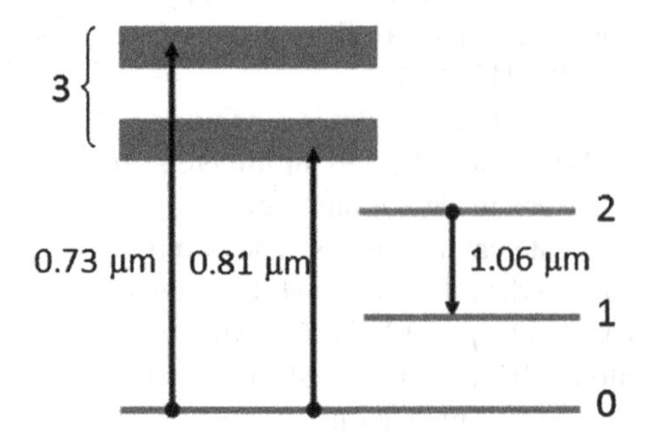

Fig. 3.14 Neodymium laser: simplified scheme of the energy levels of the ion Nd^{3+}

AlGaAs semiconductor laser, which emits at 0.81 μm, and is in resonance with one of the absorption lines of the Nd^{3+} ion. The important advantage in such a scheme is that all the pump power can be used to create the population inversion, while less than 50 % of the lamp pump power has a wavelength overlap that drives the excitation from level 0 to level 3. Another important advantage is that the collimated light from the semiconductor pump laser can be easily directed into the target laser medium. This is not the case for discharge lamps, because they emit over the whole solid angle, making it difficult to fully harness all the pump power available. Typically a Nd laser pumped by a semiconductor laser has an output power of about 30 % the input pump power.

The Nd laser is useful in a great many applications, including materials processing (drilling, welding, marking of industrial products, etc.), optical radar, and biomedical applications. It can also be used to create green laser light, through a process known as second-harmonic generation, as it will be discussed in Chap. 7.

A useful property of rare earth ions (Nd, Er, Yb, etc.) is that they can provide optical amplification even when randomly dispersed in a transparent glass matrix, instead of a highly ordered crystal. A particularly important exploitation of this property is represented by the development of fiber-optic amplifiers and oscillators, as will be discussed in Chap. 6.

Gas lasers. Shortly after the invention of the ruby laser, it was shown that it is also possible to have population inversion in a gaseous medium. Gas lasers are almost always pumped using an electric discharge inside the active medium. The gas is typically contained inside a capillary tube. If the voltage applied across the tube is sufficiently large, a fraction of the gas atoms (or molecules) becomes ionized, and the gas, which is an insulator at rest, becomes a conductor of electric current. The free electrons, accelerated by the external electric field, collide with the atoms, generating a great variety of excited states. By choosing the gas mixture and pressure correctly, along with the value of the electric current density, laser oscillation can be achieved in many gaseous media.

The first gas laser, which also was the first CW laser (Bell Telephone Laboratories, 1960), was the He-Ne laser, made using a transition between two excited levels of the neon atom, emitting infrared radiation at 1.15 μm. The capillary tube is filled with a mixture of helium and neon atoms, typically in the ratio 10 : 1, at a pressure around one hundredth of an atmosphere. The helium atoms excited by electronic collisions selectively transfer their excitation energy to neon atoms, generating population inversion on certain neon transitions. Besides operating at 1.15 μm, the He-Ne laser can emit at 0.633 and 3.39 μm. The most interesting wavelength for applications is the red emission. Commercial He-Ne lasers typically have a cavity length between 10 and 100 cm, and an output power between 1 and 50 mW. The output beam has a very stable pointing direction and a very small angular spread—the divergence angle is typically 10^{-4} rad. The laser can be frequency-stabilized, in order to provide a monochromatic beam with a linewidth of the order of 1 kHz. During a few decades following its invention, the He-Ne laser was the laser of choice for those applications requiring low power and high beam-stability, such as bar-code reading, telemetry, velocimetry, and alignment in civil and mechanical engineering. In most of these applications the He-Ne laser has now been replaced by the smaller and cheaper semiconductor laser.

An important gas laser is the carbon dioxide (CO_2) laser, emitting in the middle infrared at the wavelength $\lambda = 10.6\,\mu\text{m}$, corresponding to a transition between two excited vibrational levels of the CO_2 molecule. The active gas mixture, made of carbon dioxide, nitrogen and helium, is pumped by an electric discharge, which can be longitudinal (electric current flowing along the axis of the optical cavity) or transversal. The gas mixture is usually flowed along the laser tube in order to refresh the active medium. The structure of the CO_2 laser depends on the desired output power. The most common configuration utilizes a longitudinal electric discharge and a slow longitudinal gas flow. Cooling the lateral walls of the laser tube with a water circulation system provides the necessary heat dissipation. With a laser tube 1 m long, filled at a pressure of 2.5×10^3 Pa, a continuous output power of the order of 100 W can be produced. By replacing the slow gas flow with a supersonic flow (which also acts as an effective means of heat removal) the output power can be increased to 1 kW. A even higher output power of up to 20 kW can be reached by increasing the pressure to $\approx 1.3 \times 10^4$ Pa and using a transversal electric discharge and transversal gas flow. In some cases, a radio-frequency electric discharge (at 30–50 MHz) is used instead of a continuous discharge. This has the advantage that no electrodes inside the laser tube are needed. The CO_2 laser has a very high efficiency, typically $\eta \approx 20\,\%$. This is an extremely important property for a high-power laser used in industrial applications. Indeed the main applications of the CO_2 laser are in the area of materials processing. Surgical applications have also been tested, using the laser as a lancet to exploit the fact that biological tissues strongly absorb 10.6-μm radiation.

Another useful class of gas lasers is the excimer laser, which emits powerful pulses of ultraviolet light. In this laser a pulsed electric discharge creates excited bi-atomic molecules formed by a noble gas element (argon or krypton) and a halogen element (fluorine or chlorine). The molecule is formed because the chemical behavior of an excited noble gas atom is similar to that of an alkali atom. The excited molecules dissociate by emitting ultraviolet light. For instance, the argon-fluorine (Ar-F) excimer laser can emit 10–100 ns pulses at 193 nm, with a repetition rate of hundreds of Hz and an average output power of several watts. The excimer laser has very important applications in the fields of photolithography for microelectronics and ophthalmology for the myopia treatment.

3.10 Pulsed Lasers

Some of the most important applications of lasers involve pulsed operation. There are different methods of obtaining laser pulses. One method is to pulse the pump driving the laser emission, as it is the case for excimer lasers. Another possibility is to use an external switch that transmits the output of a CW laser for short periods of time. Naturally, the minimum pulse duration that can be obtained with this approach cannot be shorter than the switch response time, and the peak power of the pulse cannot exceed the steady-state power of the CW laser. These limitations can be overcome, as will be shown, by inserting an internal switch or modulator directly into the laser cavity.

3.10.1 Q-Switching

The technique of "Q-switching" is used to generate short and powerful light pulses from lasers that have an upper-level spontaneous decay time τ much longer than the photon decay time of the laser cavity, i.e., $\tau \gg \tau_c$.

The principle is as follows: an optical switch is inserted into the cavity, before one of the mirrors. When the pump is switched on and the switch is closed introducing extra losses, the population inversion can build up to a very high value, but does not reach the oscillator threshold. When the population inversion reaches its peak value, the switch is suddenly opened, and the loop gain of the cavity is almost instantly boosted well above the oscillation threshold. This results in an extremely rapid build-up of the photon number and a subsequent depletion of the population inversion by stimulated emission. This process converts most of the pump energy stored by the excited atoms into the emission of a short and powerful laser pulse.

The name Q-switching refers to the quality factor Q of the laser cavity, which is a figure of merit quantifying the cavity losses, and is defined as:

$$Q = \frac{\textit{angular frequency} \times \textit{energy stored}}{\textit{power dissipated}}. \tag{3.63}$$

Applying this general definition to the laser case, it is found that $Q = \omega_{21}\tau_c$. The process above described changes suddenly the state of the laser cavity from large losses (short τ_c, that is, low Q) to small losses (long τ_c, that is, high Q), hence the name Q-switching.

The transient behavior following the switch-on operation at time $t = 0$ can be derived by solving the two time-dependent rate Eqs. (3.49) and (3.50), with appropriate initial conditions. The initial number of excited atoms, $N_2(0)$, is calculated from (3.50) under the assumptions that (i) there are no photons in the cavity and (ii) the time derivative is equal to 0. This gives: $N_2(0) = W\tau$. If the initial number of photons is actually set to zero, then the system cannot dynamically evolve. However, in practice there are always some initial photons present in the cavity, due to spontaneous emission. A typical transient, obtained by choosing $Q(0) = 1$, is sketched in Fig. 3.15. The upper curve shows that the pumping process makes the population of the upper level to grow much larger than the threshold value when the optical switch is closed. The cavity Q is switched on at $t = 0$. Then the number of photons begins to grow, leading to a pulse with a characteristic half-height duration, τ_p, of the order of τ_c. As a result of the growth in photon number, the population of the upper state will decrease from its initial value to a very low final value.

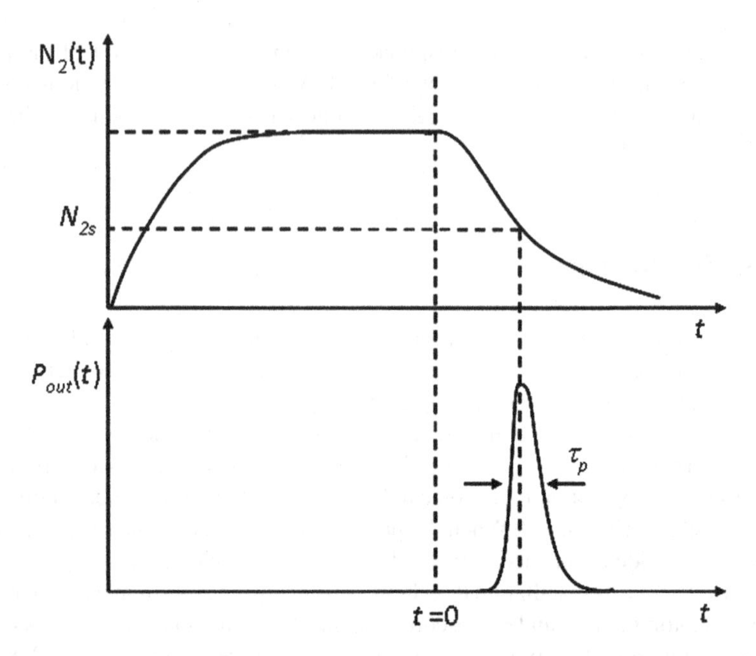

Fig. 3.15 Time evolution of the population of the upper laser level (*upper diagram*) and of the laser output power (*lower diagram*) in a Q-switched laser. The Q-switching operation is performed at $t = 0$

In order to derive the order of magnitude of the peak power, it can be assumed that the total number of photons inside the pulse coincides with the total number of excited atoms at $t = 0$, $VN_2(0)$, leading to a peak output power estimate of:

$$P_{peak} = h\nu \frac{VW\tau}{\tau_c}. \tag{3.64}$$

Recalling that the steady-state output power P_s is given by (3.55), and assuming, for simplicity, that $W \gg W_{th}$, the ratio of peak power to steady-state power in the emission is:

$$\frac{P_{peak}}{P_s} = \frac{\tau}{\tau_c}. \tag{3.65}$$

Equation (3.65) clearly shows that the Q-switching operation is only advantageous for active media having a long spontaneous decay time of their excited levels, such as solid-state and CO_2 lasers. If one takes $\tau = 300\,\mu s$ and $\tau_c = 10\,ns$ for the case of a Nd-YAG laser, then the power ratio can be up to 3×10^4. A laser designed for an average output power of 1 W can have a Q-switching pulse (sometimes called a "giant pulse" in the literature) with a peak power of 30 kW.

The optical switch used in Q-switching can be realized exploiting the electro-optic or acousto-optic effects later described in Chap. 4. Another interesting possibility, frequently utilized, is the inclusion of a saturable absorber in the optical cavity for "passive" Q-switching. The absorber can be modeled as the two-level system described in Sect. 3.4, with an intensity-dependent absorption. The absorption loss is high in the early stages of the oscillation build-up, because the slowly increasing intensity is not large enough to saturate the absorption. However, once the intensity in the cavity exceeds a certain level (the saturation level, see (3.17)), the loss quickly decreases, and the resulting effect is similar to that of a sudden opening of an optical switch.

3.10.2 Mode-Locking

Q-switching produces pulses with a duration ranging from a few nanoseconds to hundreds of nanoseconds. Another technique, known as "mode-locking", is capable of producing pulses on the femtosecond and picosecond scale. Whereas Q-switching can work equally well with lasers in single-mode or multi-mode operation, mode-locking requires a large number of longitudinal modes in the cavity.

Consider a CW laser oscillating on N longitudinal modes. The frequency difference between adjacent modes is $\Delta\nu$. For a Fabry-Perot cavity, $\Delta\nu$ is given by (3.48).

The electric field associated to the laser beam is written as:

$$E(t) = \sum_{k=1}^{N} E_k \exp[-i(\omega_k t + \phi_k)], \tag{3.66}$$

where the amplitudes E_k are real and time-independent. The laser intensity, which is proportional to $|E(t)|^2$, is the sum of the intensities of the N modes plus the sum of $N(N-1)$ interference terms, each having zero time-average. If the phases ϕ_k are random and mutually uncorrelated, then the interference terms tend to cancel each other, such that:

$$I(t) = \sum_{k=1}^{N} I_k + \Delta I(t), \tag{3.67}$$

where $\Delta I(t)$ is a small randomly fluctuating term.

The mode-locking operation consists in introducing a fixed relation among the phases of different modes, expressed by:

$$\phi_k - \phi_{k-1} = \phi_o, \tag{3.68}$$

where ϕ_o is an arbitrary constant. To simplify the algebra, it is assumed that N is an odd number and that the frequency of the central mode coincides with the central frequency ν_{21} of the gain band. Since N is odd, it can be written as: $N = 2n + 1$.

By calling m the integer number that takes all values in the interval $-n \le m \le n$, assuming equal amplitudes, $E_m = E_o$, and using (3.68), (3.66) then becomes:

$$E(t) = \sum_{m=-n}^{n} E_o \exp\{-i[(\omega_o + m\Delta\omega)t + m\phi_o]\} = A(t)E_o \exp(-i\omega_o t), \quad (3.69)$$

where

$$A(t) = \sum_{m=-n}^{n} \exp[-im(\Delta\omega t + \phi_o)]. \quad (3.70)$$

It is convenient to introduce the new temporal variable t', connected to t by the relation $\Delta\omega t + \phi_o = \Delta\omega t'$. To use t' instead of t is equivalent to shifting the time axis by the quantity $\phi_o/\Delta\omega$. The summation in the right-hand member of (3.70) is a geometric progression with common ratio $\exp(i\Delta\omega t')$. It is found:

$$A(t') = \sum_{-n}^{n} \exp(-im\Delta\omega t') = \frac{\sin[(2n+1)\Delta\omega t'/2]}{\sin(\Delta\omega t'/2)}. \quad (3.71)$$

The time-dependent intensity is derived from (3.71) by recalling that $I(t')$ is proportional to $E_o{}^2 A^2(t')$. As an example, the function $I(t')$ is plotted in Fig. 3.16, for $N = 11$ and $\Delta\nu = 100$ MHz. The intensity profile is a periodic sequence of pulses, separated by a period of:

$$T = \frac{2\pi}{\Delta\omega} = \frac{1}{\Delta\nu}. \quad (3.72)$$

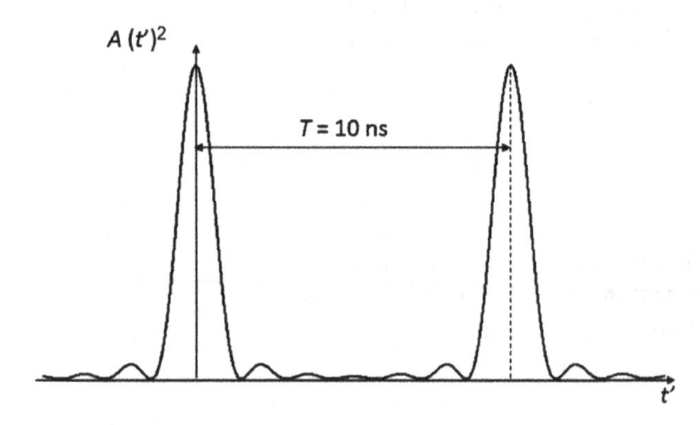

Fig. 3.16 Temporal behavior of the output intensity of a mode-locked laser in the case of an equal-amplitude mode spectrum. In this numerical example 11 modes with a frequency separation of 100 MHz are considered

The peak intensity, I_p, is obtained by calculating $A^2(t')$ in the limit of t' tending to 0. The result is:

$$I_p = N^2 I_o = N I_L, \tag{3.73}$$

where I_o is the intensity of one mode and $I_L = N I_o$ is the average laser intensity.

It is important to note that the pulsed behavior is the consequence of an interference effect, As such, mode-locking does not change the average power of the emission, it simply re-distributes the power non-uniformly in time. Therefore, the energy contained in a single pulse must be equal to that of the integrated power emitted by the CW laser over a period T. Consequently, the pulse duration τ_p can be determined from the following equality:

$$\tau_p I_p = T N I_L \quad . \tau_p = \frac{T}{N}. \tag{3.74}$$

Recalling that T is the reciprocal of $\Delta\nu$ and that $N = \Delta\nu_{21}/\Delta\nu$, the very important result is found that $\tau_p = (\Delta\nu_{21})^{-1}$, regardless of the cavity length. As an example, the Nd-YAG laser has a typical gain bandwidth of 200 GHz, and so the minimum pulse duration that can be obtained from mode-locked operation is 5 ps. For a Nd-glass laser, the bandwidth is one order-of-magnitude broader than that of a Nd-YAG laser, and so its minimum pulse duration is of the order of 500 fs. The pulse duration can be reduced by another order of magnitude by mode-locking the Ti-sapphire laser.

The Fourier transform of the emitted field, $E(\omega)$, can be described as the product of an infinite comb of Dirac delta functions with a rectangle function:

$$E(\omega) = \sum_{k \to -\infty}^{\infty} E_o rect\left(2\frac{\omega - \omega_o}{\Delta\omega_{12}}\right)\delta(\omega - \omega_k) \tag{3.75}$$

where $rect(x)$ is a function which takes value 1 if $-0.5 \leq x \leq 0.5$ and 0 if $|x| > 0.5$. Recalling that the inverse Fourier transform of the product of two functions of ω is the convolution of the two anti-transforms, and that the Fourier anti-transform of $rect$ is the sinc function, it can be verified that the anti-transform of (3.75) is precisely given by (3.69). The interesting aspect of this Fourier approach is that it allows for a general treatment to the more realistic case of unequal mode amplitudes. Indeed the typical situation would be that in which the amplitudes of E_k are distributed according to a bell-shaped function, with maximum at $k = 0$. By using a Gaussian distribution, $E(\omega)$ can be expressed as:

$$E(\omega) = \sum_{k \to -\infty}^{\infty} E_o \exp\left[\frac{(\omega_o - \omega_k)^2}{\Delta\omega_{12}{}^2}\right]\delta(\omega - \omega_k). \tag{3.76}$$

It is apparent that the anti-transform of (3.76) gives an $E(t)$ that consists in a periodic sequence of Gaussian pulses, each having a duration inversely proportional to the bandwidth. It should be noted that the secondary maxima that are present in the intensity plot of Fig. 3.16 disappear when a Gaussian distribution of mode amplitudes is used instead of the rectangle function.

Mode-locking can be achieved in various ways. One method is to internally modulate the losses of the laser at a frequency which is equal to the intermode frequency spacing $\Delta\nu$. With this approach the laser behavior can be intuitively understood as a self-adjusting process: if losses are minimal at times t_o, $t_o + T$, etc., in order to maximize the efficiency the laser will prefer to emit pulses at t_o, $t_o + T$, etc., instead of operating at constant intensity. The loss modulation can be obtained by inserting into the laser cavity an electro-optic or acousto-optic modulator.

The description of the N-mode laser as a collection of N independent oscillators is a rather rough approximation. In practice, even without any external intervention, the modes are interacting because they are driven by emission from the same active atoms. Generally speaking, the existence of interactions among the modes tends to favor the process of mode-locking. What prevents the laser from spontaneously operating in a mode-locked regime is the fact that real longitudinal modes are not exactly equally spaced in frequency. The optical path appearing in (3.47) contains the index of refraction of the active medium, which is a frequency-dependent quantity. As a consequence, (3.47) should be viewed as an implicit expression for the resonance frequencies of the laser cavity, producing a spacing that is slightly different for each pair of adjacent modes. Only if the modulation depth exceeds a threshold value, the amplitude (or phase) modulation forces the modes to become equally spaced and locks the mode phases.

Mode-locking can also be achieved by passive methods. A frequently used approach is to insert a saturable absorber in the cavity. Initially, the multimode laser has random intensity fluctuations, but the saturable absorber selectively transmits the fluctuations of high intensity, while absorbing those of low intensity. After many passes through the cavity, only a single high-intensity pulse is formed. In this scheme, it is essential to use a saturable absorber with a recovery time much shorter than the period T. Semiconductor saturable-absorber mirrors (SESAMs) can also provide passive mode-locking. Furthermore, as it will be explained in Chap. 7, spontaneous mode-locking can be achieved in solid-state lasers by exploiting the nonlinear optical behavior of the active medium. This method is called Kerr-lens mode locking.

Ultrashort pulses of high peak power can be obtained by simultaneously operating in the same laser cavity mode-locking and Q-switching. In such a situation the obtained mode-locked train has an envelope that follows the time evolution of the Q-switching pulse. As an example, Fig. 3.17 illustrates the result of an experiment in which a passively mode-locked diode-pumped $Nd\text{-}YVO_4$ laser is Q-switched by an acousto-optic switch. The output train is made up of about 20 mode-locked pulses, separated by a period of 5.6 ns corresponding to the cavity round trip time. The measured width of the pulses was about 9 ps.

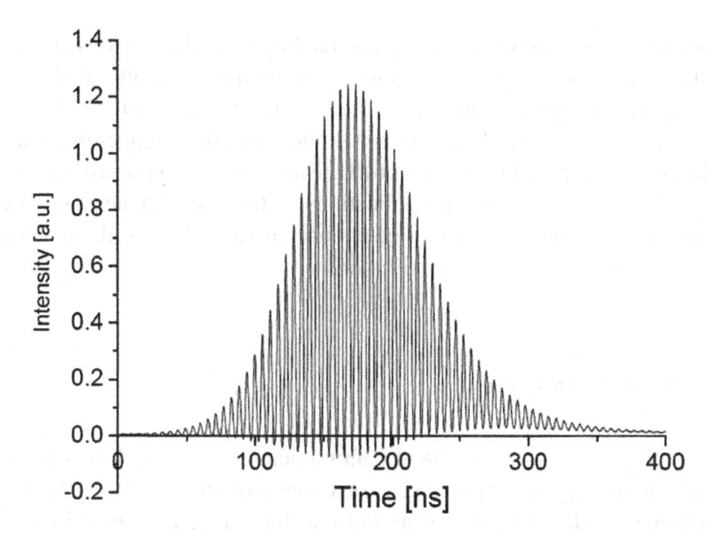

Fig. 3.17 Output intensity of a simultaneously mode-locked and Q-switched $Nd\text{-}YVO_4$ laser

3.11 Properties of Laser Light

Laser light differs from the light emitted by conventional sources in many aspects. The most relevant properties of laser emission are reviewed below.

3.11.1 Directionality

Laser radiation is characterized by a high degree of directionality. While traditional light sources emit in all directions and with a wide frequency range, the laser (as an oscillator) has a built-in selection mechanism for frequency and direction. The origin of the directionality can be intuitively understood, if one considers that only rays traveling back-and forth between the cavity mirrors, along the cavity axis, are strongly amplified by the active medium. As discussed in Sect. 3.8, the spatial properties of laser beams can be described in terms of Gaussian waves. If the cavity supports several transverse modes, then the fundamental mode TEM_{00}, can be selected by placing an aperture within the cavity. The TEM_{00} laser emission has an angular spread θ_o typically in the range $10^{-4} - 10^{-3}$ radians.

The beam can be further collimated by a telescope consisting of a pair of two converging lenses sharing a common focus. If the focal length of the first lens is f_1 and that of the second lens is $f_2 > f_1$, then the beam diameter is increased by the factor f_2/f_1, and, consequently, the beam divergence is decreased by the same factor.

In many applications, such as writing and reading optical memories, a laser beam is focused to a very small spot size. Using a lens with focal length f, the spot size in the focal plane is approximated by $w_f \approx \theta_o f$. Taking $\theta_o = 10^{-4}$ and $f = 1$ cm, it is found that $w_f = 1\,\mu$m. This shows that the small divergence of a laser beam enables the concentration of the output power to an area having the size of the optical wavelength. This is useful for many applications. In the case of materials processing, the intensity at focus can be easily large enough to strongly heat the material, and locally produce melting or even vaporization.

3.11.2 Monochromaticity

There are applications, such as optical communications, in which it is essential to have a single-frequency optical source. This requires suppression not only of higher-order transverse modes, but also of all-but-one longitudinal modes. In almost all cases, the bandwidth of the frequency-dependent gain profile $\Delta\nu_{21}$ is much larger than the frequency spacing between adjacent longitudinal modes, $\Delta\nu$, and so the laser has a natural tendency to operate with multiple modes. Single-mode behavior can be enforced by using compound cavities or, more effectively, by using a frequency-selective feedback mechanism. An example of this approach will be given in Chap. 5, where single-mode semiconductor lasers will be discussed.

The emission frequency of a single-mode laser, ν_L, always has fluctuations originated by external disturbances. Since ν_L is linearly dependent on the optical path L' between the two laser mirrors, the frequency stability of the laser depends on the stability of L'. As shown in Sect. 2.4.4, to shift by $\Delta\nu$ the mode frequency, it is sufficient to change the mirror separation by half wavelength. This means that small variations in the mirror separation can induce large changes of the laser emission wavelength. Therefore, in order to minimize the wavelength fluctuations of the single-mode laser, the cavity length must be kept fixed by rigid construction, with isolation from mechanical vibrations, and also with temperature control. In a stable single-mode laser the linewidth $\delta\nu_L$ of the emitted radiation can be of the order of 1 MHz, and even lower values of 1 kHz can be achieved using active frequency-stabilization methods.

A useful feature of some types of lasers is the possibility to tune the emission wavelength without changing the active medium, but simply acting on an optical component inside the cavity.

It should be noted that, in principle at least, one can filter both the direction and frequency of light emitted by a conventional source, in order to achieve a divergence and bandwidth comparable to a laser source. However, the resulting beam would be not only extremely weak, but also characterized by large amplitude and phase fluctuations.

3.11.3 Spectrum of Laser Pulses

The light emitted by conventional optical sources contains many frequency components. Each component has a random phase, uncorrelated with that of the other components. In the laser case, it is possible to have a multi-frequency emission in which the phases of different frequency components are mutually correlated, as it has been shown in Sect. 3.10.2. This is very important because it allows for the possibility of optical signals having a pre-determined time-dependent (and space-dependent) amplitude and phase, analogous to what can be done in the range of radio-frequencies or microwaves with an electronic function generator.

In this section the relation between pulse duration and spectral width is derived by considering a specific pulse shape, that of the time-dependent field amplitude described by a Gaussian function. Assuming that the intensity peak is at $t = 0$, the electric field can be expressed as:

$$E(t) = A_o \exp\left(-\frac{t^2}{2\tau_i{}^2}\right)\exp(-i\omega_o t), \tag{3.77}$$

where ω_o is the central frequency, and τ_i is the time delay at which the intensity is reduced by a factor $1/e$ with respect to the peak value. The pulse duration τ_p, defined as the width at half-height of the intensity peak, is related to τ_i as follows:

$$\tau_p = 2\sqrt{ln2}\,\tau_i = 1.67\tau_i. \tag{3.78}$$

Recalling the definite integral:

$$\int_0^\infty \exp(-a^2t^2)dt = \frac{\sqrt{\pi}}{2a}, \tag{3.79}$$

the Fourier transform of (3.77) can be immediately calculated, obtaining:

$$E(\omega) = \sqrt{2\pi}\,\tau_i A_o \exp\left(-\frac{\tau_i{}^2(\omega-\omega_o)^2}{2}\right). \tag{3.80}$$

Note that $E(\omega)$ is a real function, which means that all the frequency components have the same phase. This type of optical pulse is known as a "transform-limited" pulse. Equation (3.80) shows that the power spectrum of the Gaussian pulse, $S_p(\omega) \propto |E(\omega)|^2$, is also a Gaussian function. The Gaussian pulse, together with its power spectrum, is illustrated in Fig. 3.18. The width at half-height of the power spectrum, $\Delta\nu_p$, satisfies the relation:

$$\Delta\nu_p\tau_p = \frac{2\,ln2}{\pi} = 0.441. \tag{3.81}$$

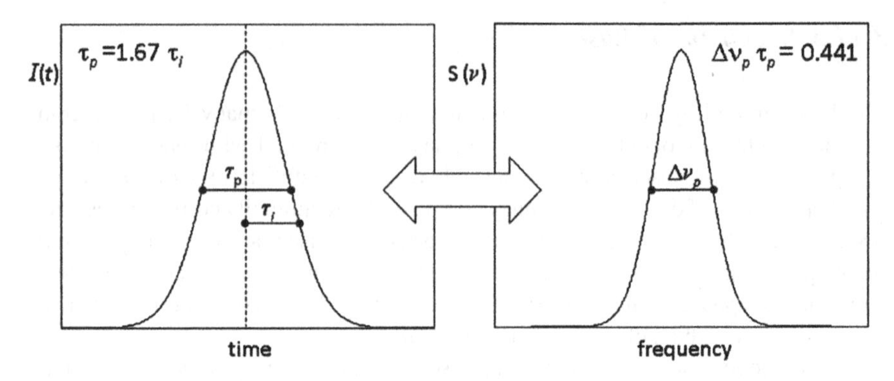

Fig. 3.18 The Gaussian pulse and its power spectrum

It is a general property of transform-limited light pulses that the bandwidth-duration product is a number of the order of 1. The precise value of the product depends on the shape of the pulse. In the more general case of pulses which are not transform-limited, (3.81) gives only the lower bound of the bandwidth-duration product. To consider an extreme situation, a short pulse of solar light can be created by using a fast switch on a Sun beam. This pulse would have the spectrum of Sun light, regardless of its duration, so that the bandwidth-duration product would be larger than 0.441 by many orders of magnitude.

Problems

3.1 What is the momentum of a 10^{18}-Hz X-ray photon?

3.2 If 0.5 % of the light intensity incident into a medium is absorbed per millimeter, what fraction is transmitted if the medium is 10 cm long? Calculate the absorption coefficient α.

3.3 Calculate the photon decay time τ_c for a ring laser cavity having length 30 cm and mirror reflectances $R_1 = 0.95$ and $R_2 = 0.85$.

3.4 An Nd-YAG laser having a Fabry-Perot cavity of length 30 cm and an active medium of length 5 cm is operated in the mode-locking regime. Assuming that the index of refraction of the active medium is $n = 2$ and that the gain bandwidth is 200 GHz, calculate the repetition period T and the duration τ_p of the emitted pulses.

3.5 Considering that the gain bandwidth of a He-Ne laser is 1.5 GHz, which is the maximum Fabry-Perot cavity length that ensures single longitudinal mode operation?

3.6 Calculate the mirror reflectance required to sustain oscillation in a symmetric Fabry-Perot laser cavity, assuming that the active medium has a small-signal gain coefficient of 1 m^{-1} and is 0.1 m long.

3.7 Consider a Fabry-Perot cavity Nd-YAG laser, in which the Nd-YAG crystal has length $L = 2\,\text{cm}$ and mirror 1 is totally reflecting. Assuming that the population inversion inside the active medium is $N_2 - N_1 = 1 \times 10^{18}\,\text{cm}^{-3}$, the stimulated emission cross-section is $\sigma_{21} = 3.5 \times 10^{-19}\,\text{cm}^2$, the loop gain is equal to 2, determine the reflectance of mirror 2.

Chapter 4
Modulators

Abstract The application of an external field can modify the optical properties of a material. By controlling the strength and frequency of the external field, the amplitude, phase, or polarization of the optical beam can be controlled and modulated. The most important modulation methods are those using electric fields, but both acoustic fields and magnetic fields can be exploited. This chapter describes linear and quadratic electro-optic effects, and explains the most common modulation and switching schemes. Modulation methods based on liquid crystals are also briefly discussed. The final part of the chapter is devoted to the acousto-optic effect and its applications to amplitude modulation and to beam deflection.

4.1 Linear Electro-Optic Effect

The electro-optic effect is the change in the index of refraction due to an applied electric field. If the refractive index change is proportional to the applied field, then the effect is known as Pockels effect. If the index change is quadratically-dependent on the applied field, then the effect is known as Kerr effect. As it will be seen, the Pockels effect can exist only in crystalline materials, and is absent in amorphous materials, such as glasses and fluids. In contrast, the Kerr effect is present in all classes of materials.

In general, crystals are anisotropic media, in which **D** and **P** are related to the field **E** by a second-rank tensor, as was seen in Chap. 2. In particular:

$$P_i(\omega) = \varepsilon_o \sum_{j=1}^{3} \chi_{ij} E_j(\omega), \tag{4.1}$$

where the indices i, j can take the values 1, 2, 3. As usual, $1 = x$, $2 = y$, $3 = z$. The electric susceptibility tensor χ_{ij} is connected to ε_{ij} by the relations $\varepsilon_{ii} = 1 + \chi_{ii}$, and, for $i \neq j$, $\varepsilon_{ij} = \chi_{ij}$. The two tensors have the same diagonal system of coordinates.

The electro-optic effect modifies the relation between **P** and **E**, by introducing additional terms dependent on an external field $\mathbf{E}'$. In the case of the Pockels effect, the additional terms are linear with respect to the external field components:

V. Degiorgio and I. Cristiani, *Photonics*, Undergraduate Lecture Notes in Physics,
DOI 10.1007/978-3-319-20627-1_4

$$P_i(\omega) = \varepsilon_o \sum_{j,k=1}^{3} (\chi_{ij} + \chi_{ijk} E'_k) E_j(\omega). \tag{4.2}$$

Equation (4.2) is nonlinear because it contains terms proportional to the product of two electric fields. A general description of nonlinear interactions will be given in Chap. 7. In the case under discussion here, it is important to keep in mind that **E** is the electric field associated with an optical wave, while **E**′ is a field generated by an electronic device. Thus the oscillation frequency of **E** is several orders of magnitude larger than that of **E**′. The external field acts as a parameter modifying the electric susceptibility of the medium, and, consequently, its refractive index.

In a centrosymmetric medium all the components of χ_{ijk} vanish. In fact, the presence of non-zero components would induce the existence of terms cubic in the electric field inside the electromagnetic energy density. Cubic terms change sign when the variable changes sign, but, in a centrosymmetric medium, the energy density cannot depend on the sign of the electric field.

The tensor χ_{ijk} has $3^3 = 27$ components, but, since it is symmetric with respect to the exchange of i with j, only 18 components are independent. For all crystal classes, symmetry considerations can be used to predict which components of χ_{ijk} are independent and non-zero.

In the scientific literature the name electro-optic tensor designs r_{ijk}, which is connected to χ_{ijk} by the relation:

$$\chi_{ijk} = -(1 + \chi_{ii})(1 + \chi_{jj}) r_{ijk} \tag{4.3}$$

A single index n can be introduced to replace the ij pair, using the following convention: $n = 1, 2, 3, 4, 5, 6$ corresponds to $ij = 11, 22, 33, 23, 13, 12$, respectively. For instance, in this contracted notation, r_{123} becomes r_{63} and r_{333} becomes r_{33}. The values of the refractive indices and of the non-zero electro-optic coefficients of some typical crystals are reported in Table 4.1.

Table 4.1 Properties of some electro-optic crystals at $\lambda = 633$ nm

Crystal	Class	Indices of refraction	r (10^{-12} m/V)
$LiNbO_3$	3m	$n_o = 2.272,\ n_e = 2.187$	$r_{33} = 30.8,\ r_{13} = 8.6$
			$r_{22} = 3.4,\ r_{51} = 28$
$KNbO_3$	mm^2	$n_1 = 2.279,\ n_2 = 2.329$	$r_{33} = 25,\ r_{13} = 10$
		$n_3 = 2.167$	
$LiTaO_3$	3m	$n_o = 2.175,\ n_e = 2.180$	$r_{33} = 33,\ r_{13} = 8$
KDP	$\bar{4}$2m	$n_o = 1.51,\ n_e = 1.47$	$r_{41} = r_{52} = 8.6,\ r_{63} = 10.6$
ADP	$\bar{4}$2m	$n_o = 1.52,\ n_e = 1.48$	$r_{41} = r_{52} = 28,\ r_{63} = 8.5$

Lithium niobate ($LiNbO_3$), potassium niobate ($KNbO_3$), lithium tantalate ($LiTaO_3$), potassium di-hydrogenated phosphate (KH_2PO_4, usually designed as KDP), ammonium di-hydrogenated phosphate ($NH_4H_2PO_4$, usually designed as ADP)

Crystals that have a symmetry supporting the Pockels effect also exhibit piezoelectric effects. This means that they expand and contract under the application of an external electric field. Piezoelectric effects are usually negligible for frequencies of the applied field larger than some MHz.

Up to frequencies of the order of 1 THz, the induced polarization contains contributions from the displacements of both ions and electrons. Above 1 THz, there are only electronic contributions and coefficients are proportionally smaller. The values reported in Table 4.1 refer to the frequency range between 1 MHz and 1 THz.

From a general point of view the wave propagation inside the electro-optic crystal should be treated by inserting (4.2) into the wave equation (2.7). Since, as it has been seen in Chap. 2, the propagation characteristics are fully described by means of the index ellipsoid, the effect of the electric field is expressed most conveniently in terms of changes of the index ellipsoid. In presence of $\mathbf{E}'$, (2.126) is modified as follows:

$$\begin{aligned}\left(\frac{1}{n_1^2}+a_1\right)x^2+\left(\frac{1}{n_2^2}+a_2\right)y^2+\left(\frac{1}{n_3^2}+a_3\right)z^2\\+2a_4yz+2a_5xz+2a_6xy=1,\end{aligned} \tag{4.4}$$

where the coefficients a_n are given by:

$$a_n=\sum_{k=1}^{3} r_{nk}E'_k \tag{4.5}$$

As an example, considering KDP, which has only three non-zero components of r_{nk}, (4.4) becomes:

$$\frac{x^2+y^2}{n_o^2}+\frac{z^2}{n_e^2}+2r_{41}E'_1yz+2r_{52}E'_2xz+2r_{63}E'_3yz=1 \tag{4.6}$$

The effect of the additional terms is that of modifying the direction of the dielectric axes and the length of the semi-axes. The modulator configurations described in the following sections will deal with rather simple situations in which $\mathbf{E}'$ is directed along the optical axis z and the light beam propagates along a principal axis.

4.1.1 Phase Modulation

The phase of an optical wave can be electro-optically modulated. For this type of application, the optical wave is linearly polarized in a direction that is parallel to one of the principal axes of the crystal in presence of the external field, so that its polarization state does not change during propagation. In order to illustrate the method two specific situations are below discussed in some detail.

Modulation based on r_{33}. Lithium niobate is a crystal often used in electro-optic modulation. It has a high electro-optic coefficient r_{33}, as shown in Table 4.1. The configuration adopted to access the r_{33} term is having the light beam traveling along x, linearly polarized along the optical axis z (extraordinary wave), and the external electric field $\mathbf{E}'$ parallel to z. In this configuration the extraordinary wave propagates without changing its state of polarization, whether $\mathbf{E}'$ is present or not. Using (4.2), it is found that the only non-zero component of $\mathbf{P}$ is that which lies along z:

$$P_3 = \varepsilon_o(\chi_{33} + \chi_{333}E_3')E_3 = \varepsilon_o(\chi_{33} - n_e^4 r_{33}E_3')E_3 \tag{4.7}$$

Equation (4.7) indicates that the extraordinary index of refraction, $n_e = \sqrt{1+\chi_{33}}$, is modified by the presence of $\mathbf{E}'$, and becomes:

$$\begin{aligned} n_e' &= \sqrt{1+\chi_{33} - n_e^4 r_{33}E_3'} = \sqrt{1+\chi_{33}}\sqrt{1 - \frac{n_e^4 r_{33}E_3'}{1+\chi_{33}}} \\ &\approx n_e - \frac{r_{33}n_e^3 E_3'}{2}, \end{aligned} \tag{4.8}$$

where, taking into account that $n_e^4 r_{33}E_3'$ is usually much smaller than $1+\chi_{33}$, the approximate expression is obtained by a truncated series expansion of the square root term.

Calling L the crystal length, the phase delay caused by the Pockels effect is given by:

$$\Delta\phi = 2\pi(n_e' - n_e)\frac{L}{\lambda} = -2\pi\frac{r_{33}n_e^3 E_3'}{2}\frac{L}{\lambda}. \tag{4.9}$$

If L is a few centimeters and λ is around 1 μm, then the ratio of L/λ is of the order of 10^4. This means that a refractive index variation of just 10^{-4} can give a phase delay $\Delta\phi$ of the order of π radians.

A phase modulator can be realized by using a lithium niobate crystal cut as a parallelepiped along the principal axes. A voltage difference V across the crystal thickness d induces an electric field $E' = V/d$ directed along z. The wave propagates without changing polarization, and acquires a phase delay $\Delta\phi$. Taking $E_o\cos(\omega_o t)$ as the electric field of the incident wave, and assuming, for simplicity, that the applied voltage oscillates with an angular frequency ω_m, such that $V(t) = V_o\sin(\omega_m t)$, the electric field of the wave that travels for a length L inside the crystal is:

$$E(t) = E_o\cos\left\{\omega_o t - \frac{2\pi}{\lambda}\left[n_e - \frac{r_{33}n_e^3}{2d}V_o\sin(\omega_m t)\right]L\right\}. \tag{4.10}$$

Equation (4.10) describes a wave at frequency ω_o, phase modulated at the frequency ω_m.

Modulation based on $\mathbf{r_{63}}$. If an external field directed along the optical axis is applied to a KDP-like crystal, then the **P** components, as derived from (4.2), are given by:

$$
\begin{aligned}
P_1 &= \varepsilon_o(\chi_{11}E_1 + \chi_{123}E'_3E_2) \\
P_2 &= \varepsilon_o(\chi_{11}E_2 + \chi_{123}E'_3E_1) \\
P_3 &= \varepsilon_o\chi_{33}E_3,
\end{aligned}
\tag{4.11}
$$

where it has been taken into account that $E'_1 = E'_2 = 0$ and $\chi_{123} = \chi_{213}$. The effective electric susceptibility is therefore described by the matrix:

$$
\begin{pmatrix} \chi_{11} & \chi_{123}E'_3 & 0 \\ \chi_{123}E'_3 & \chi_{11} & 0 \\ 0 & 0 & \chi_{33} \end{pmatrix}.
\tag{4.12}
$$

The presence of E'_3 generates off-diagonal terms which couple P_1 with E_2 and P_2 with E_1. In order to describe wave propagation inside the crystal the matrix (4.12) must be diagonalized, by introducing a new set of principal axes x', y', z', and new diagonal terms χ'_{11}, χ'_{22}, χ'_{33}. Noting that the relation between P_3 and E_3 is not modified by the presence of E'_3, it can be shown that z' must coincide with z, the plane x'-y' coincides with x-y, and $\chi'_{33} = \chi_{33}$. Since the expressions for P_1 and P_2 are symmetrical with respect to the exchange of E_1 with E_2, the angle between x' and x must be 45°. The result of the diagonalization is:

$$
\begin{aligned}
\chi'_{11} &= \chi_{11} - \chi_{123}E'_3 \\
\chi'_{22} &= \chi_{11} + \chi_{123}E'_3.
\end{aligned}
\tag{4.13}
$$

Equation (4.13) show that the rotation symmetry around the z axis is lost, because $\chi'_{11} \neq \chi'_{22}$. The new refractive indices are:

$$
\begin{aligned}
n'_1 &= \sqrt{1+\chi'_{11}} \approx n_o\left(1 + \frac{n_o^2 r_{63}E'_3}{2}\right) \\
n'_2 &= \sqrt{1+\chi'_{22}} \approx n_o\left(1 - \frac{n_o^2 r_{63}E'_3}{2}\right),
\end{aligned}
\tag{4.14}
$$

where χ_{123} has been substituted by $-n_o^4 r_{63}$, and the approximation $\chi_{123}E'_3 \ll 1+\chi_{11}$ has been made.

The scheme of a phase modulator exploiting the properties of the KDP crystal is shown in Fig. 4.1. The crystal is a parallelepiped cut along the axes x', y', z, with $\mathbf{E}'$ directed along z, the optical wave travels along z and is linearly polarized in the x' direction. Since the external voltage is applied along the propagation direction, this configuration requires transparent electrodes, which can be realized by depositing thin semiconductor layers on the crystal faces. The incident wave propagates without changing polarization, and acquires a phase delay determined by the value of n'_1.

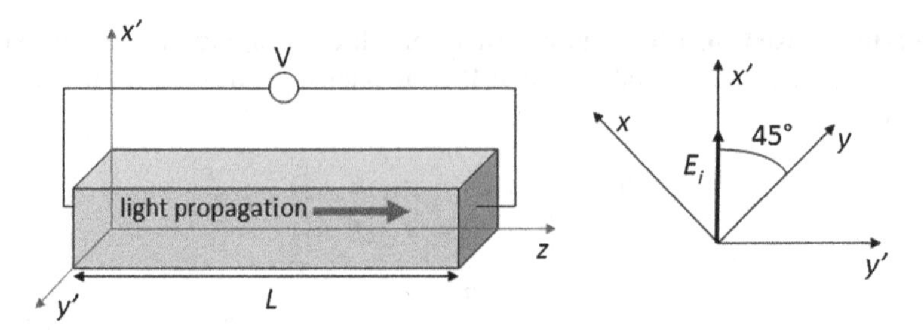

Fig. 4.1 Phase modulator based on r_{63}. The incident optical wave propagates along the optical axis, z, and is linearly polarized along x'. The external field is applied along z

Taking $E_o\cos(\omega_o t)$ as the electric field of the incident wave, and assuming again a sinusoidal applied voltage, the electric field of the wave traveling for a length L inside the crystal is:

$$E_{x'}(t) = E_o\cos\left\{\omega_o t - \frac{2\pi}{\lambda}\left[n_o + \frac{r_{63}n_o^3}{2L}V_o\sin(\omega_m t)\right]L\right\}. \tag{4.15}$$

Equation (4.15) has the same form as (4.10), such that the two schemes are conceptually equivalent. In practice, the first scheme can work with much lower applied voltages, provided the aspect ratio d/L is made very small. A more detailed comparison will be made after treating the amplitude modulation.

4.1.2 Amplitude Modulation

Phase modulation can be converted into amplitude modulation by using an interferometric method or by exploiting a birefringence variation.

Interferometric modulator. The scheme of an amplitude modulator based on a Mach-Zehnder interferometer is shown in Fig. 4.2. M_1 and M_2 are totally reflecting mirrors, the beam splitter D_1 divides the input beam in two identical parts, and the

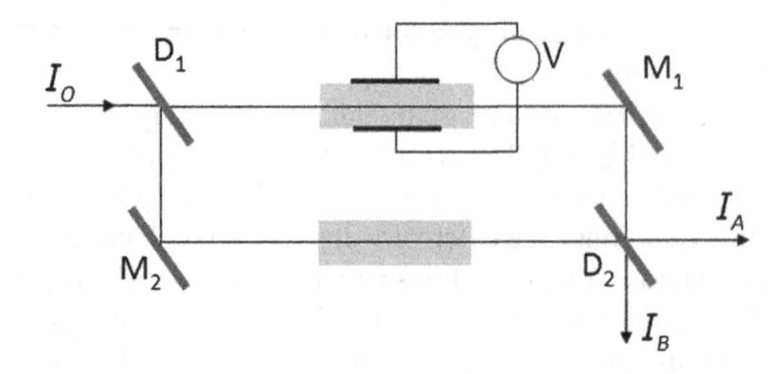

Fig. 4.2 Amplitude modulator based on the Mach-Zehnder interferometer

beam splitter D_2 recombines the two beams at the two outputs A and B. Two identical lithium niobate crystals are inserted in the upper and lower branches of the interferometer. The optical beams propagate through the crystals in a direction perpendicular to the optical axis and are linearly polarized along the optical axis. Qualitatively the behavior is as follows. If the upper and lower optical paths are identical, then there is constructive interference at output A and destructive interference at output B. If all the components are lossless, then the optical intensity at A, I_A, coincides with the input intensity, I_o, and I_B is zero. By applying an electric field to the upper crystal, the index of refraction is changed, such that the wave traveling in the upper branch will acquire an additional phase delay. As a consequence, the interferometer will be unbalanced, and some part of the input intensity will appear at output B.

The electric field of the wave at the output A (or B), E_A (E_B), is the sum of the contributions coming from the upper and lower branch of the interferometer. Taking $E_o\cos(\omega_o t)$ as the electric field of the incident wave, and considering all the phase delays experienced by the beams propagating in the upper and lower branch of the interferometer, the electric fields E_A and E_B of the two output beams are expressed as:

$$
\begin{aligned}
E_A &= \frac{E_o}{2}[\cos(\omega_o t - \phi_1) + \cos(\omega_o t - \phi_2)] \\
E_B &= \frac{E_o}{2}[\cos(\omega_o t - \phi_3) + \cos(\omega_o t - \phi_4)]
\end{aligned}
\tag{4.16}
$$

Calling ϕ_τ and ϕ_ρ the phase delays in transmission and reflection due to the beam splitters D_1 and D_2, and ϕ_M the phase delay due to reflection at M_1 and M_2, the phase delays of (4.16) are expressed as:

$$
\begin{aligned}
\phi_1 &= \phi_\tau + 2\pi L_{up}/\lambda + \phi_\rho + \phi_M \\
\phi_2 &= \phi_\rho + 2\pi L_{down}/\lambda + \phi_\tau + \phi_M \\
\phi_3 &= \phi_\tau + 2\pi L_{up}/\lambda + \phi_\tau + \phi_M \\
\phi_4 &= \phi_\rho + 2\pi L_{down}/\lambda + \phi_\rho + \phi_M .
\end{aligned}
\tag{4.17}
$$

The upper and lower optical paths are given by: $L_{up} = n_e' L + L_o$, $L_{down} = n_e L + L_o$, respectively, where L is the length of the two crystals and L_o is the path in air. Taking into account that ϕ_τ and ϕ_ρ are related by (2.58), it is found that: $\phi_1 - \phi_2 = \Delta\phi$, and $\phi_3 - \phi_4 = \Delta\phi - \pi$. $\Delta\phi$ is the phase delay caused by the Pockels effect, and is given by (4.9).

The intensity $I_A(I_B)$ at output $A(B)$ is proportional to the square of $E_A(E_B)$. Recalling that:

$$
\cos\alpha + \cos\beta = 2\cos\left(\frac{\alpha+\beta}{2}\right)\cos\left(\frac{\alpha-\beta}{2}\right)
$$

it is found that:

$$
I_A = I_o\cos^2\left(\frac{\phi_1 - \phi_2}{2}\right) = I_o\cos^2\left(\frac{\Delta\phi}{2}\right), \tag{4.18}
$$

and

$$I_B = I_o \cos^2\left(\frac{\phi_3 - \phi_4}{2}\right) = I_o \sin^2\left(\frac{\Delta\phi}{2}\right). \tag{4.19}$$

Equations (4.18) and (4.19) show that, if $\Delta\phi = \pm\pi$, then the whole input intensity is transferred to output B. The applied voltage needed for a phase shift of π, known as the half-wave voltage, is given by:

$$V_\pi = \frac{\lambda d}{r_{33} L n_e{}^3}. \tag{4.20}$$

Since the transmittance of the Mach-Zehnder interferometer, observed at output B, ranges from 0 to 1 when the applied voltage is changed from 0 to V_π, the interferometer can be used as an electro-optic switch.

If the applied voltage is a continuously varying signal, for instance the sinusoidal function $V(t) = V_m \sin(\omega_m t)$, then the device behaves as an amplitude modulator. However, the output intensity is not linearly dependent on the applied voltage. In order to obtain a linear relation, at least in the small signal regime, the working region should be shifted around the inflexion point of the curve I_B/I_o versus V, as shown in Fig. 4.3. This can be achieved by adding a DC bias voltage $V_{\pi/2} = V_\pi/2$, which introduces a fixed phase shift $\pi/2$. The total applied voltage becomes: $V = V_{\pi/2} + V_m \sin(\omega_m t)$. By using (4.19), the ratio I_B/I_o takes the expression:

$$\frac{I_B}{I_o} = \sin^2\left[\frac{\pi}{4} + \frac{\pi V_m \sin(\omega_m t)}{2V_\pi}\right]. \tag{4.21}$$

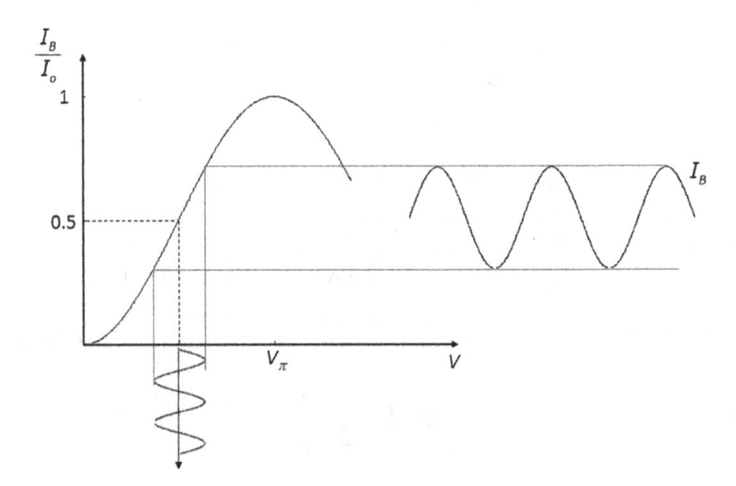

Fig. 4.3 Analog modulation based on the Mach-Zehnder interferometer

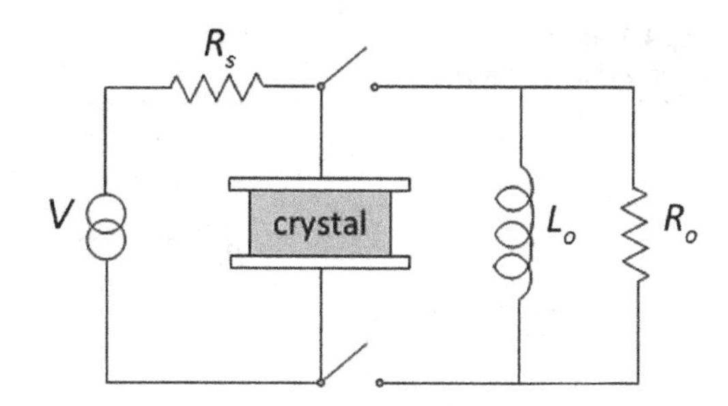

Fig. 4.4 Equivalent electric circuit of the crystal with applied voltage

If $V_m/V_\pi \ll 1$, (4.21) can be approximated as:

$$\frac{I_B}{I_o} = \frac{1}{2}\left\{1 + 2\sin\left[\frac{\pi V_m \sin(\omega_m t)}{2V_\pi}\right]\right\} \approx \frac{1}{2}\left[1 + \frac{\pi V_m \sin(\omega_m t)}{V_\pi}\right], \tag{4.22}$$

obtaining a linear relation between applied voltage and output intensity. The amplitude modulation used in optical communications operates around the inflexion point of the curve I_B versus V.

The intrinsic response time of the Pockels effect can be smaller than 10^{-11} s. The limits to the speed of the device come from electrical capacitive effects: the crystal behaves as a capacitor with capacity C, charged by a voltage generator with internal resistance R_s (see Fig. 4.4). Therefore, if the modulation frequency ω_m is larger than $(R_s C)^{-1}$, then most of the applied voltage will be dropped across R_s instead of across the crystal. Considering the problem in the time domain, the response time is given by $R_s C$. Recalling that $C = \varepsilon_o \varepsilon_r A/d$, where A is the surface area, a quick calculation shows that the typical response time is of the order of 1 ns.

A method to reduce the response time is to put a coil with inductance L_o and a resistor R_o in parallel to the crystal, such that ${\omega_m}^2 = (L_o C)^{-1}$ and $R_o \gg R_s$. At resonance, the impedance of the parallel $R_o L_o C$ is equal to R_o, and so the applied voltage is dropped almost completely across the crystal. The price to pay in this configuration is that the device can operate only in a band of modulation frequencies centered at ω_m with a width $\Delta\omega = (R_o C)^{-1}$.

The modulation induced by the applied field is effective only if the transit time of the optical wave through the crystal is smaller than the modulation period $T_m = 2\pi/\omega_m$. Indeed, instead of using (4.9), $\Delta\phi$ should be more generally calculated as:

$$\Delta\phi = -\frac{2\pi r_{33} {n_e}^3}{2\lambda}\int_0^L E_3' dz = -\frac{\omega r_{33} {n_e}^2}{2}\int_t^{t+\tau_d} E_3'(t')dt', \tag{4.23}$$

where $\tau_d = L n_e/c$ is the transit time through the crystal, and $E_3'(t')$ is the instantaneously applied field. In the right-hand member of (4.23), the spatial integration has been substituted by a temporal integration by putting $dt' = n_e dz/c$. If $E_3' = E_o' \sin(\omega_m t)$, $\Delta\phi$ is given by:

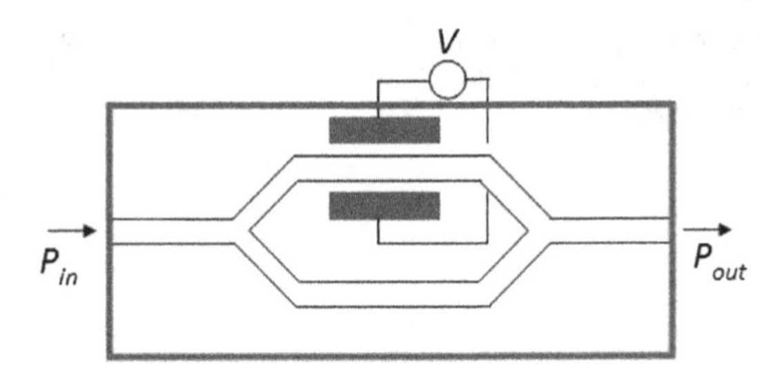

Fig. 4.5 Integrated Mach-Zehnder modulator

$$\Delta\phi = -\frac{\omega r_{33} {n_e}^2 \tau_d E_o'}{2} \left[\frac{\sin(\omega_m \tau_d)}{\omega_m \tau_d} \right]. \tag{4.24}$$

In the limit $\omega_m \tau_d \rightarrow 0$, (4.24) coincides with (4.9). Assuming that the upper limit to the modulation frequency is determined by the condition $\omega_m \tau_d = \pi/2$, it is found that the maximum value is: $\nu_{max} = c/(4Ln_e)$. By using a lithium niobate crystal long 1 cm, $\nu_{max} = 3\,\text{GHz}$.

The lithium niobate modulators utilized inside optical communication systems are integrated devices, in which the light beams travel inside embedded strip waveguides fabricated by in-diffusing in the lithium niobate substrate a material (usually titanium) that increases the extraordinary refractive index. A scheme of the integrated modulator is shown in Fig. 4.5. The width of the waveguide is very small, typically $d = 5\ \mu\text{m}$, so that the half-wave voltage is reduced to about 1 V, as calculated from (4.20) by assuming a waveguide length of a few centimeters. Light is coupled into and out of the modulator by using optical fibers. The output capacitance is strongly reduced in comparison with the bulk device, and so the response time is much faster. Transit-time effects can be, in principle, eliminated by using the electrodes as a transmission line in which the velocity of the traveling electrical signal matches that of the light beam. Modulation speeds of several tens of GHz can be achieved with this approach.

Modulation by birefringence variation. In order to perform an amplitude modulation based on r_{63}, $\mathbf{E}'$ is directed along z, and the optical wave travels along z, with a linear polarization along x, as shown in Fig. 4.6. Since in this case the wave is not polarized along a principal axis, its propagation has to be treated by separately considering the two components of the electric field that are parallel to x' and y' and travel inside the crystal with different velocities. If the electric field of the incident wave is $E_x = E_o\cos(\omega_o t - kz)$, then the two components of the wave exiting the crystal are:

$$E_{x'} = \frac{E_o}{\sqrt{2}}\cos(\omega_o t - \phi_{x'})\ ; \quad E_{y'} = \frac{E_o}{\sqrt{2}}\cos(\omega_o t - \phi_{y'}). \tag{4.25}$$

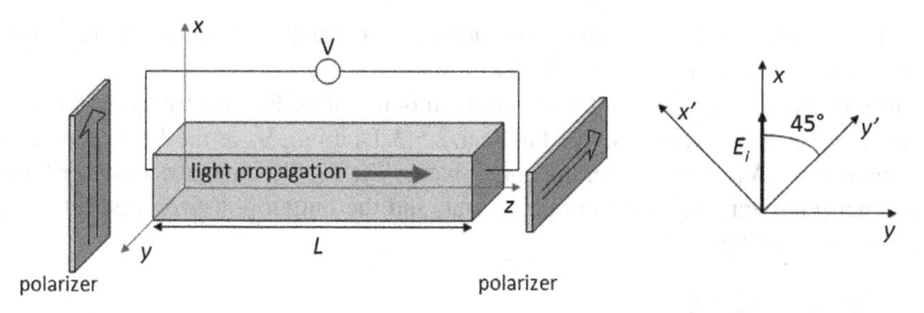

Fig. 4.6 Amplitude modulator based on r_{63}. The input wave is linearly polarized along x. The axis of the output polarizer is along y

The relative phase delay is:

$$\Delta\phi = \phi_{x'} - \phi_{y'} = \frac{2\pi n_o^3 r_{63} V}{\lambda}, \tag{4.26}$$

where $V = LE'_3$ is the applied voltage. The output wave is elliptically polarized, that is, a modulation of the polarization is produced. In order to transform the device into an amplitude modulator, it is sufficient to add a polarizer with the optical axis parallel to y, as shown in Fig. 4.6. The electric field of the wave transmitted by the polarizer is the sum of the projections of $E_{x'}$ and $E_{y'}$ on the y axis:

$$\begin{aligned} E_y &= (E_o/2)[\cos(\omega_o t - \phi_{y'}) - \cos(\omega_o t - \phi_{x'})] \\ &= E_o \sin(\omega_o t - \phi_o)\sin(\Delta\phi/2), \end{aligned} \tag{4.27}$$

where $\phi_o = (\phi_{x'} + \phi_{y'})/2$. The ratio between output and input intensity is:

$$\frac{I_u}{I_o} = \sin^2\left(\frac{\Delta\phi}{2}\right) = \sin^2\left(\frac{\pi V}{2V_\pi}\right), \tag{4.28}$$

where

$$V_\pi = \frac{\lambda}{2r_{63}n_o^3} \tag{4.29}$$

is the half-wave voltage for this configuration.

Equation (4.28) is identical to (4.19), therefore the two amplitude modulators are equivalent in principle. However, by comparing (4.29) with (4.20), a very important difference appears: if the electric field $\mathbf{E}'$ is applied transversally to the direction of wave propagation, then V_π becomes proportional to the aspect ratio d/L. As a consequence, if a small aspect ratio is used for the transverse modulation, as in the case of the integrated modulator of Fig. 4.5, then V_π can be of the order of 1 V. In contrast, the longitudinal modulation scheme of Fig. 4.6, taking also into account that the electro-optic coefficient of KDP is much smaller than that of lithium niobate, has a half-wave voltage of several kilovolts.

It is instructive to discuss the amplitude modulator of Fig. 4.6 by using the formalism of Jones matrices described in Sect. 2.5.2. Defining $\mathbf{V_o}$ as the Jones vector of the input wave, $\mathbf{V_u}$ the vector of the output wave, $\mathbf{T_1}$, $\mathbf{T_2}$, $\mathbf{T_3}$, the matrices describing the input polarizer, the electro-optic crystal, and the output polarizer, respectively, then $\mathbf{V_u}$ is given by:

$$\begin{aligned}\mathbf{V_u} &= \mathbf{T_3 T_2 T_1 V_o} \\ &\begin{pmatrix}0 & 0\\ 0 & 1\end{pmatrix}\begin{pmatrix}(\exp(-i\Delta\phi)+1)/2 & (\exp(-i\Delta\phi)-1)/2\\ (\exp(-i\Delta\phi)-1)/2 & (\exp(-i\Delta\phi)+1)/2\end{pmatrix}\begin{pmatrix}1 & 0\\ 0 & 0\end{pmatrix}\begin{pmatrix}1\\ 0\end{pmatrix} \\ &= \begin{pmatrix}0\\ \left[\exp(-i\Delta\phi)-1\right]/2\end{pmatrix}.\end{aligned} \tag{4.30}$$

The ratio between output and input intensity is simply given by:

$$\frac{I_u}{I_o} = \left|\frac{\exp(-i\Delta\phi)-1}{2}\right|^2 = \frac{1-\cos(\Delta\phi)}{2} = \sin^2\!\left(\frac{\Delta\phi}{2}\right), \tag{4.31}$$

which coincides with (4.28).

In order to operate the modulator in a linear region, a quarter-wave plate, having x' and y' as principal axes, can be inserted after the crystal. Using the x and y axes as the reference system, the matrix describing the plate is:

$$\mathbf{T_2'} = \begin{pmatrix}(\exp(-i\pi/2)+1)/2 & (\exp(-i\pi/2)-1)/2\\ (\exp(-i\pi/2)-1)/2 & (\exp(-i\pi/2)+1)/2\end{pmatrix}. \tag{4.32}$$

The output signal becomes:

$$\mathbf{V_u} = \mathbf{T_3 T_2' T_2 T_1 V_i} = \begin{pmatrix}0\\ (\exp(-i(\Delta\phi+\pi/2)-1)/2\end{pmatrix}, \tag{4.33}$$

and the ratio I_u/I_o is given by:

$$\frac{I_u}{I_o} = \frac{1-\cos(\Delta\phi+\pi/2)}{2} = \sin^2\!\left(\frac{\Delta\phi}{2}+\frac{\pi}{4}\right). \tag{4.34}$$

The insertion of the quarter-wave plate has the effect of shifting the operating point of the modulator to the inflexion point of the curve I_u/I_o versus V. Note that in the case of the interferometric amplitude modulator described in the previous section, the same effect was obtained by applying a bias voltage equal to $V_{\pi/2}$ (see Fig. 4.3).

A transverse modulation scheme could also be applied to the KDP modulator, as shown in Fig. 4.7, where $\mathbf{E}'$ is parallel to z, and the light beam propagates in the y' direction, with a linear polarization at 45° with z. The phase delay $\Delta\phi$ between the x' and z components is given by:

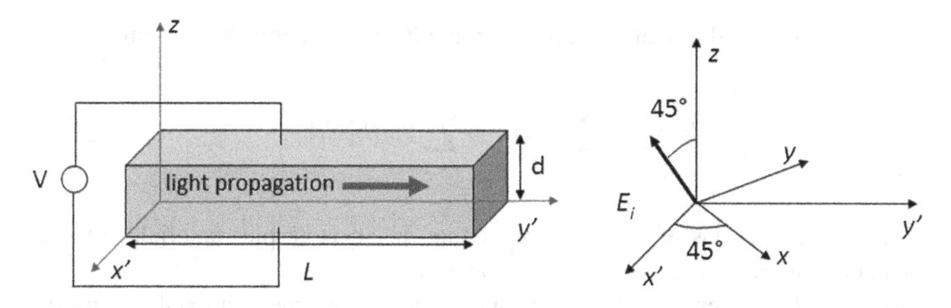

Fig. 4.7 Amplitude modulator based on r_{63} in a transverse scheme. The external electric field is applied along z and the light beam propagates along y'

$$\Delta\phi = \phi_{x'} - \phi_z = \frac{2\pi L}{\lambda}\left(n_o - n_e + \frac{n_o^3 r_{63} V}{2d}\right), \tag{4.35}$$

where d is the crystal thickness. Equation (4.35) shows that $\Delta\phi$ is non-zero even in absence of E'. This is obvious, because the light beam is not propagating along the optical axis. However, the presence of the voltage-independent term complicates the utilization of the transverse configuration.

The electro-optic methods described in this section are also utilized for the generation of ultrashort laser pulses. In the case of a Q-switched laser, a fast optical switch can be realized by inserting into the cavity an electro-optic shutter that blocks the laser beam when voltage is applied to the crystal and is suddenly opened by removing the voltage. Electro-optic modulators can be used to drive a laser in the mode-locking regime by intra-cavity modulation at a frequency $\omega_m \approx 2\pi\,\Delta\nu$, where $\Delta\nu$ is the frequency difference between two adjacent longitudinal modes of the laser cavity.

4.2 Quadratic Electro-Optic Effect

Centrosymmetric materials may exhibit a quadratic electro-optic effect, i.e. a refractive index change proportional to the square of the applied electric field:

$$\Delta n = n_K E'^2, \tag{4.36}$$

where n_K, measured in m^2/V^2, is the Kerr constant. It should be mentioned that many authors define differently the Kerr constant, by writing $\Delta n = \lambda n'_K E'^2$, where λ is the wavelength of the modulated light beam and the constant n'_K is measured in $\mathrm{m/V^2}$.

In general terms, the relation between **P** and **E** takes the following form:

$$P_i = \varepsilon_o \left(\sum_j \chi_{ij} E_j + \sum_{j,k,l} \chi_{ijkl} E'_k E'_l E_j \right) \tag{4.37}$$

The tensor χ_{ijkl} has $3^4 = 81$ components. In an isotropic material, only the components of the type χ_{iiii} or χ_{iikk} are non-zero.

Because of the Kerr effect, an isotropic material becomes uniaxial, taking the direction of the applied electric field as the optical axis. Assume that χ is the electric susceptibility and $n = \sqrt{1+\chi}$ the refractive index of the isotropic material, and that $\mathbf{E}'$ is parallel to z. For a wave traveling along y and linearly polarized along z the extraordinary index of refraction is:

$$n_e = n + n_K^e E'^2 \tag{4.38}$$

where n_2^e is proportional to the component χ_{3333}. For the wave traveling along y, but polarized along x, the ordinary refractive index is:

$$n_o = n + n_K^o E'^2, \tag{4.39}$$

where n_2^o is proportional to χ_{1133}. Therefore the two Kerr constants are generally different.

In solid media the Kerr effect is mainly connected with a displacement of electron clouds, whereas in fluid media important contributions may come from the orientation of anisotropic molecules. The electro-optic effects of electronic origin are usually small, but have very short response times. In contrast, orientational Kerr effects can be large, but have correspondingly long response times.

In absence of an applied field a liquid consisting of elongated molecules behaves as an isotropic medium because the molecular orientations are completely random. When an electric field is applied, the molecules tend to orient parallel to each other, and the medium becomes uniaxial. If $\mathbf{E}'$ is not very large, then the first-order theory predicts that the change in refractive index is proportional to E'^2, and that $n_K^o = -n_K^e/2$. For large $\mathbf{E}'$, the molecules become all parallel to the field, and the medium birefringence attains a saturation value.

In principle the Kerr effect can be utilized for electro-optic modulation by using schemes similar to those of Pockels modulators. In practice, the most important applications involve liquid crystals, and are discussed in the next section.

A very interesting aspect of the Kerr effect is that, being quadratic, it can also be effective when the applied field is oscillating at very large frequencies, such as the optical frequencies. This is known as the optical Kerr effect, and will be discussed in Chap. 7.

4.2.1 Liquid Crystal Modulators

A particularly large quadratic electro-optic effect is shown by liquid crystals, with very important applications in displays or in other situations not requiring fast response times. In normal cases a solid material, when heated to its melting point, becomes a liquid, and undergoes a phase transition from an ordered state to a disordered state. Liquid crystals are a particular class of materials that present one (or more) intermediate states between the crystalline solid and the isotropic liquid. They are fluid-like, yet the arrangement of molecules exhibits structural order. A characteristic of liquid crystals is the rod-like shape of their molecules. Within the liquid crystal family, there are three distinct molecular structures: nematic, cholesteric, and smectic, as shown in Fig. 4.8. Only the first two structures are of importance in modulators and displays. In the nematic phase, the elongated molecules are aligned parallel to each other, but they are free to move relative to each other so that the phase has liquid properties. In the cholesteric phase, the material is made up from a large number of planes each having a nematic-like structure, but with each plane showing a progressive change in the molecular axis direction from the one below. The molecular axis directions thus display a helical twist through the material.

The nematic temperature range of a single compound is usually limited to a few tens of degrees. For instance, the compound known as $5CB$ is crystalline below 24 °C, is nematic between 24° and 35.3 °C, and becomes a liquid above 35.3 °C. For display applications a wide nematic range is highly desirable. To widen the nematic range, mixtures of several components are commonly used.

Liquid crystals are optically anisotropic: the nematic phase is birefringent and the cholesteric phase presents rotatory power. A light beam traveling across a nematic liquid crystal layer in a direction perpendicular to the molecular orientation behaves as an extraordinary wave if linearly polarized along the molecular axis. If an applied electric field forces the molecules to align parallel to the propagation direction of the light beam, then the wave becomes an ordinary wave. Therefore the effect of the electric field is that to change the phase delay experienced by the light beam, from $2\pi n_e L/\lambda$ to $2\pi n_o L/\lambda$, where L is the layer thickness. Since the liquid crystal birefringence can be quite large, typically $n_e - n_o \approx 0.1$, a difference in phase delay equal to π can be obtained with a layer thickness of just a few microns. It is therefore

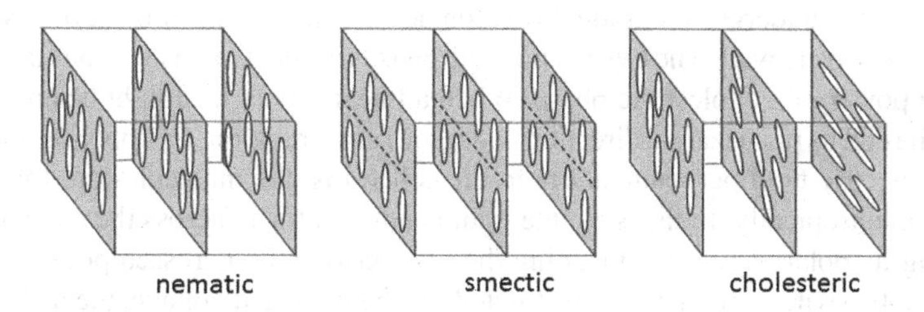

Fig. 4.8 Liquid crystals

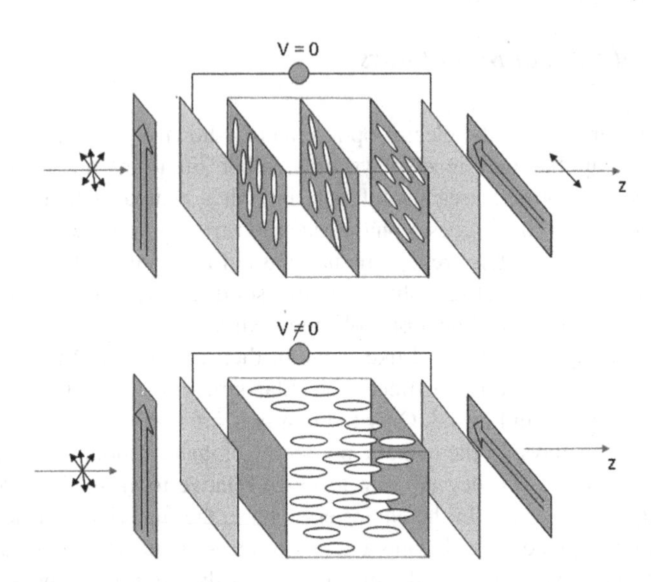

Fig. 4.9 Optical switch based on a twisted nematic liquid crystal cell. The switch is open if there is no applied voltage. The light beam is blocked if the liquid crystal molecules are forced to align parallel to z by a sufficiently large voltage

possible to make an electro-optic switch or modulator using a very thin liquid crystal film.

When a nematic liquid crystal material comes into contact with a solid surface, the molecules can become aligned either perpendicular to the surface (homeotropic ordering) or parallel to the surface (homogeneous ordering), depending on the treatment of the surface. In the case of homogeneous ordering, the orientation direction of the molecules is selected by rubbing the surface along a particular direction before it comes into contact with the liquid crystal. The molecules will take up an orientation parallel to the direction of rubbing.

A typical example of electro-optic effects in liquid crystals is given by the cell using a "twisted nematic liquid crystal", as shown in Fig. 4.9. The liquid crystal layer is closed inside a glass cell, in which opposing walls are treated to produce a homogeneous ordering with orientations at right angles to each other. Thus the molecular axis undergoes a gradual rotation across the cell, from 0° at one wall to 90° at the opposite wall. The twisted nematic phase obtained in this way has the same rotatory power of a cholesteric phase. When a linearly polarized light beam travels across the cell its polarization direction undergoes a 90° rotation. By applying a strong enough electric field perpendicularly to the cell walls, the molecules are forced to align homeotropically. In this state the light beam will travel across the cell without changing its polarization. By inserting the cell between two crossed polarizers, an electro-optic switch can then be constructed. With no applied voltage, the light beam is transmitted through the second polarizer. When voltage is applied the direction of polarization of the light traversing the cell is not rotated and hence cannot pass through the second polarizer. The device operates with an applied voltage of a few volts, but has long response times, since the electro-optic effect is based on molecular rotations. The typical application is in the field of displays, as it will be discussed in Chap. 8.

4.3 Acousto-Optic Effect

The acousto-optic effect is the change in the refractive index of a medium due to the mechanical strains associated with the passage of an acoustic wave inside the medium. The refractive index changes are caused by the photoelastic effect which occurs in all materials under a mechanical stress.

Consider the case of a monochromatic plane wave of frequency ν_i and wave vector $\mathbf{k_i}$, traveling in a medium in which an acoustic wave of frequency ν_s and wave vector $\mathbf{k_s}$ is present. Qualitatively the interaction between the light wave and the acoustic wave can be described in the following way. The acoustic wave induces a periodic variation of the refractive index, which constitutes a three-dimensional phase grating, as shown in Fig. 4.10. The periodicity is given by the acoustic wavelength $\lambda_s = u_s/\nu_s$, where u_s is the sound velocity. It is assumed that $\mathbf{k_s}$ is parallel to the z axis, and that $\mathbf{k_i}$ lies in the y-z plane, forming an angle θ with the y-axis, as shown in Fig. 4.11. Considering the acoustic wavefronts as partially reflecting surfaces, a diffracted beam can be formed only if there is constructive interference among the reflections coming from all the reflecting surfaces, similarly to the case of the multilayer mirror treated in Sect. 2.2.4. It is clear from this approach that the diffracted beam, if it exists, will have a wave vector $\mathbf{k_d}$ forming an angle θ with the y-axis.

The condition for constructive interference is found to be:

$$\sin\theta = \frac{k_s}{2k_i} = \frac{\lambda_i}{2n\lambda_s}, \tag{4.40}$$

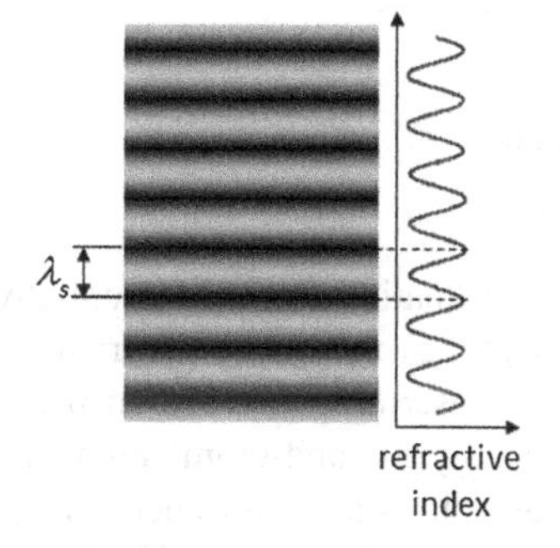

Fig. 4.10 An acoustic wave induces a traveling periodic variation of the refractive index

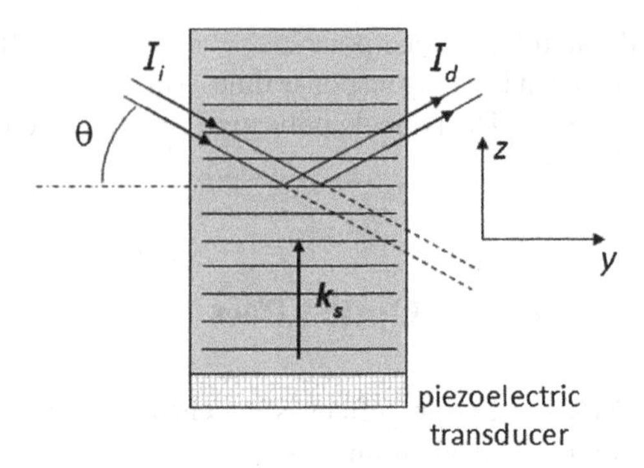

Fig. 4.11 Acousto-optic diffraction

where n is the index of refraction of the medium in which the waves are propagating. Once λ_i and $\mathbf{k_s}$ are fixed, a diffracted beam is found only if the direction of the incident beam satisfies (4.40).

In practice, the diffraction angle is small, as shown by the following example. Consider a visible light beam at $\lambda_i = 600\,\text{nm}$, and a sound wave of frequency $\nu_s = 100\,\text{MHz}$. Assuming a sound velocity inside the medium of $u_s = 3 \times 10^3$ m/s, one finds $\lambda_s = u_s/\nu_s = 30\,\mu\text{m}$. If $n = 1.5$, (4.40) gives $\theta = 0.4°$. This means that the diffracted beam forms an angle $\alpha = 2\theta = 0.8°$ with the incident beam.

Since the acoustic wavefronts are traveling at velocity $\mathbf{u_s}$, one should expect that the frequency of the diffracted beam be Doppler shifted with respect to that of the incident beam by the amount:

$$\Delta\nu_D = \frac{(\mathbf{k_d} - \mathbf{k_i}) \cdot \mathbf{u_s}}{2\pi} = \nu_s. \tag{4.41}$$

The mathematical treatment of the interaction between optical wave and acoustic wave, both taken as plane monochromatic waves, is sketched-out in the following section. A very important result is that the frequency ν_d and the wave vector $\mathbf{k_d}$ of the diffracted beam must satisfy the following conditions:

$$\nu_d = \nu_i \pm \nu_s, \tag{4.42}$$

and

$$\mathbf{k_d} = \mathbf{k_i} \pm \mathbf{k_s}. \tag{4.43}$$

Equation (4.42) is essentially a confirmation of (4.41). Equation (4.43) is graphically expressed by the triangles in Fig. 4.12. The two equations can be interpreted as describing a collision between two quasi-particles, the incident photon, with energy $h\nu_i$ and momentum $(h/2\pi)\mathbf{k_i}$, and the quantum of acoustic energy, known as "phonon", with energy $h\nu_s$ and momentum $(h/2\pi)\mathbf{k_s}$. The choice of the sign inside (4.42) and (4.43) depends on the direction of the energy and momentum exchange between optical and acoustic waves. The plus sign indicates that the phonon

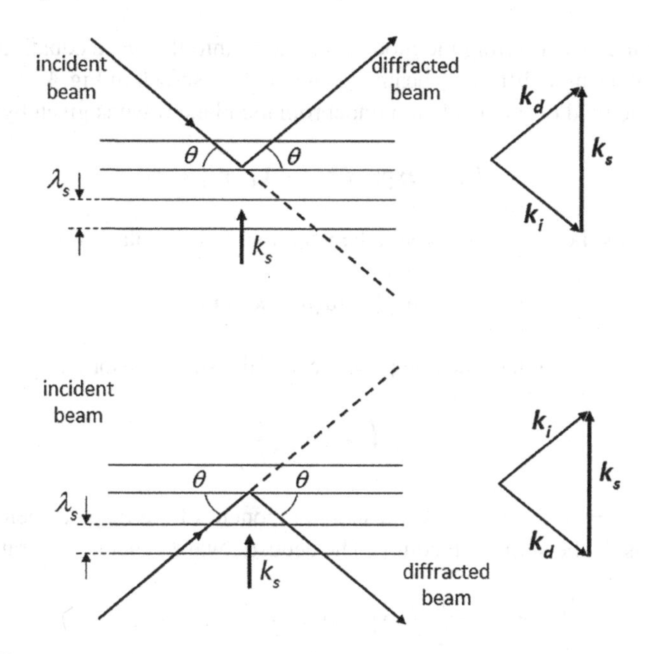

Fig. 4.12 Conservation of wave vectors in two cases. Upper part: a phonon is absorbed. Lower part: a phonon is generated

is absorbed, and its energy and momentum are added to the diffracted photon. Viceversa, the minus sign applies when a phonon is generated. In the situation illustrated in Fig. 4.11, the acoustic wave transfers energy and momentum to the optical wave, and so the plus sign is chosen. Of course, if the acoustic wave is a stationary wave, then both signs are present at the same time. In the following examples, it will be assumed that the plus sign applies.

Since the frequency of the acoustic wave is many orders of magnitude smaller than the optical frequency, ν_d is very close to ν_i. Therefore it is a very good approximation to assume that the modulus of $\mathbf{k_d}$ is equal to that of $\mathbf{k_i}$. As a consequence, the triangles of Fig. 4.12 can be seen as isosceles, so that it is immediately verified that $\sin\theta = k_s/(2k_i)$, as it has been already anticipated by the relation (4.40).

4.3.1 Acousto-Optic Modulation

The treatment of the acousto-optic interaction develops similarly to that of the electro-optic effect. The procedure is here outlined without detailed calculations, only reporting the final result.

The acoustic wave modifies the relation between **E** and **P** by introducing a strain-dependent term. The propagation of the light wave in the presence of the acoustic field is described by inserting the modified relation into the wave equation, and by assuming that a single diffracted beam is generated, as shown in Fig. 4.11.

The electric field of the incident monochromatic plane wave is given by:

$$E_i = (E_{io}/2)\exp[-i(\omega_i t - \mathbf{k_i} \cdot \mathbf{r})] + c.c.$$

The diffracted beam is also taken to be a monochromatic plane wave. Its electric field is:

$$E_d = (E_{do}/2)\exp[-i(\omega_d t - \mathbf{k_d} \cdot \mathbf{r})] + c.c.$$

The general definition of the component S_{kl} of the strain tensor is:

$$S_{kl} = \frac{1}{2}\left(\frac{\partial u_k}{\partial x_l} + \frac{\partial u_l}{\partial x_k}\right),$$

where $u_k(\mathbf{r})$ ($k = 1, 2, 3$) is the Cartesian component of the displacement $\mathbf{u}(\mathbf{r})$ of the point $\mathbf{r}$ inside the strained medium. The acoustic wave generates a strain wave:

$$S_{kl}(\mathbf{r}, t) = \frac{1}{2} S^o_{kl} \exp[-i(\omega_s t - k_s \cdot \mathbf{r})] + c.c. \tag{4.44}$$

In presence of the strain wave, the relation between electric field and electric polarization becomes:

$$P_i = \varepsilon_o \sum_{jkl=1}^{3} \left(\chi_{ij} + f_{ijkl} S_{kl}\right) E_j, \tag{4.45}$$

where f_{ijkl} is the acousto-optic tensor. It is usually more convenient to express the equations in terms of the photoelastic tensor, p_{ijkl}, related to f_{ijkl} by:

$$f_{ijkl} = -(1 + \chi_{ii})(1 + \chi_{jj}) p_{ijkl}.$$

Limiting the treatment to isotropic media, the relation between the acousto-optic and the photoelastic parameters simplifies to: $f = -n^4 p$. As a consequence, instead of (4.45), the following expression can be used:

$$\mathbf{P} = \varepsilon_o[\chi - n^4 p S(\mathbf{r}, t)]\mathbf{E}, \tag{4.46}$$

where $S(\mathbf{r}, t) = S_o \exp[-i(\omega_s t - k_s \cdot \mathbf{r})]$ is now the amplitude of the strain wave. Calculations are made in the stationary regime, such that the field amplitudes E_{io} and E_{do} are taken as time-independent quantities. As the incident wave is propagating inside the medium, its amplitude E_{io} decreases because some of the incident power is transferred to the diffracted beam. Consequently, the amplitude of E_{do} grows from its initial value of zero. In principle, the amplitude of the acoustic wave is also changed by the interaction. However, the variation of acoustic power is negligible in all the cases of practical interest.

By imposing the initial conditions of $E_{io}(0) = E_o$ and $E_{do}(0) = 0$, the treatment gives the following dependence of the field amplitudes on the propagation distance y:

$$E_{io}(y) = E_o \cos(hy\cos\theta)$$
$$E_{do}(y) = iE_o \sin(hy\cos\theta), \tag{4.47}$$

where:

$$h = \frac{\pi p n^3}{2\lambda_i} S_o. \tag{4.48}$$

Taking L as the path-length inside the medium, the final result is that the ratio between diffracted and incident intensity is given by:

$$\frac{I_d}{I_o} = \sin^2\left(\frac{\pi n^3 p S_o L}{2\lambda_i}\right) \tag{4.49}$$

The oscillating behavior of the diffracted optical intensity versus L is explained as follows. The incident wave transfers power to the diffracted wave until it is completely extinguished. However, in turn, the diffracted wave generates a secondary diffracted wave that exactly coincides with the incident wave. The result is that there is a periodic exchange of power between the two waves during propagation, with the sum of the two powers remaining constant in an ideal lossless material.

By using the amplitude S_o of the acoustic wave as a modulating signal, acousto-optic modulation can be achieved. Since the experimental control parameter is the intensity of the acoustic wave, I_s, it is better to express the diffracted intensity as a function of I_s. Recalling that:

$$S_o = \sqrt{\frac{2I_s}{\rho_m {u_s}^3}}, \tag{4.50}$$

where ρ_m is the medium density, and further defining a figure of merit, M, that summarizes the relevant material properties:

$$M = \frac{n^6 p^2}{\rho_m {u_s}^3}, \tag{4.51}$$

Equation (4.49) becomes:

$$\frac{I_d}{I_o} = \sin^2\left(\frac{\pi L}{\sqrt{2}\,\lambda_i}\sqrt{MI_s}\right) \tag{4.52}$$

As an example, if water is chosen as the acousto-optic material, the appropriate material values are: $n = 1.33$, $p = 0.31$, $u_s = 1500\,\text{m/s}$, $\rho_m = 1000\,\text{kg/m}^3$. The figure of merit is then calculated as: $M = 1.58 \times 10^{-13}\,\text{m}^2/\text{W}$. Considering an acoustic power of 1 W over an area of 1 mm × 1 mm, and taking $L = 1$ mm with $\lambda_i = 0.6\,\mu\text{m}$, it is found that $I_d/I_o = 0.97$. This example shows that comparatively strong optical effects may be obtained using moderate acoustic power and short propagation lengths.

One of the most commonly used materials in commercial acousto-optic modulators is tellurium dioxide, TeO_2, which has a wide transparency window (400–4000 nm) and $M = 34 \times 10^{-15}\,\text{m}^2/\text{W}$.

In order to easily separate the diffracted beam from the incident beam the diffraction angle θ should not be too small. In any case it should be somewhat larger than the divergence angle of the optical beam. Since θ is proportional to the frequency of the acoustic wave, as shown by (4.40), this criterion fixes the lower limit to the acoustic wave frequency.

An acousto-optic cell can act as an optical switch for the Q-switching operation. In the presence of an acoustic wave, a significant portion of the optical beam traveling inside the laser cavity is diffracted out of the cavity, increasing the cavity losses and keeping the laser below threshold. Once the acoustic wave is shut off by suddenly putting the voltage across the piezoelectric transducer to zero, laser action can start and the giant pulse is generated. Acoustic cells are also used in mode-locked lasers: instead of having an acoustic traveling wave, an acoustic stationary wave is produced. If the amplitude of the stationary wave oscillates at the frequency $\Delta\nu = c/(2L')$, then the cell behaves as an amplitude modulator that forces the laser to operate in the mode-locking regime.

4.3.2 Acousto-Optic Deflection

The acousto-optic effect can be used to generate a deflected beam in a controlled way, without recurring to moving mirrors or prisms. Deflectors are useful in a variety of applications, such as writing and reading operations with laser beams. Furthermore, since the deflection angle depends on λ_i, the acousto-optic cell behaves similarly to a dispersive prism, and can be used as a frequency filter.

As shown in the preceding section, the angle between the diffracted beam and the incident beam is $\alpha = 2\theta$, where θ is given by (4.40). If θ is small, then $\sin\theta \simeq \theta$, hence:

$$\alpha = 2\theta = \frac{\lambda_i}{n\lambda_s} = \frac{\lambda_i}{nu_s}\nu_s \tag{4.53}$$

Equation (4.53) suggests that a scanning of the deflection angle can be achieved by simply making a sweep of the acoustic frequency. If the sweep covers an interval $\Delta\nu_s$, then the corresponding variation of the deflection angle is:

$$\Delta\alpha = \frac{\lambda_i}{nu_s}\Delta\nu_s. \tag{4.54}$$

The important figure here is the number of distinguishable angular positions, N, determined by the ratio between $\Delta\alpha$ and the angular divergence of the laser beam. $\theta_o = \lambda_i/(\pi w)$, w being the radius of the Gaussian beam. The ratio is:

$$N = \frac{\Delta\alpha}{\theta_o} = \Delta\nu_s\tau, \tag{4.55}$$

where $\tau = w/u_s$ is the transit time of the acoustic wave across the light beam. As an example, if $u_s = 1500$ m/s (water), $w = 5$ mm, ν_s is swept from 80 to 125 MHz ($\Delta\nu_s = 45$ MHz), one finds: $N = 150$.

There is an important conceptual point that should be clarified. If both $\mathbf{k_i}$ and the direction of $\mathbf{k_s}$ are kept fixed, then the condition (4.43), which graphically corresponds to the closure of the triangle in Fig. 4.12, does not give any degree of freedom: diffraction is obtained at one acoustic frequency only. This would imply that one would not get any diffracted beam during a sweep of the acoustic frequency, except at that instant in which the acoustic wave vector satisfies the equality (4.43). How can this problem be overcome? The key to the solution lies in the observation that the acoustic wave is not truly a plane wave. It is transversally limited to a finite cross-section, and this results in a spread in the direction of $\mathbf{k_s}$, as depicted in Fig. 4.13. Therefore, as ν_s is changed, it may still be possible to close the triangle of wave vectors by exploiting a component of the acoustic wave that has a different direction.

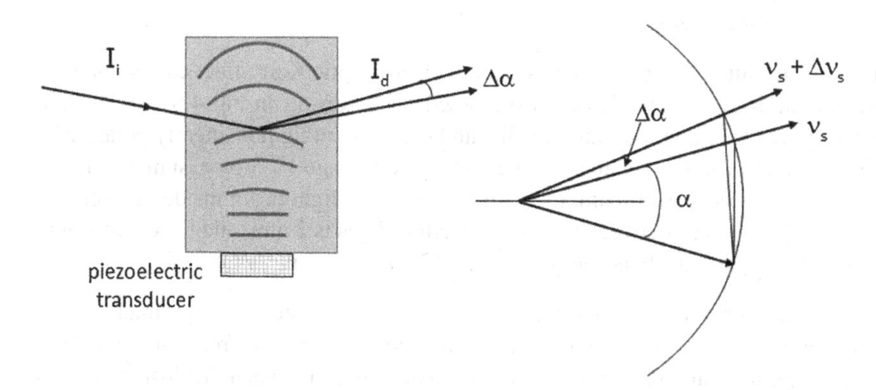

Fig. 4.13 Acousto-optic deflection

Problems

4.1 Consider the electro-optic amplitude modulator of Fig. 4.2, using a lithium niobate crystal. The operating wavelength is $\lambda = 1530$ nm, the electro-optic crystal is lithium niobate, the crystal length is $L = 6$ cm, the crystal thickness is $d = 40$ μm, and the output intensity I_B is equal to zero when no voltage is applied to the crystal. (i) Calculate the value of the applied voltage V_π required to obtain 100 % transmission at input B; (ii) Calculate the upper limit to the modulation frequency, as determined by the condition that the transit time of the light beam through the crystal be one fifth of the modulation period.

4.2 Consider the electro-optic amplitude modulator of Fig. 4.2, using a lithium tantalate crystal. The operating wavelength is $\lambda = 1550$ nm, the crystal length is $L = 4$ cm, the crystal thickness is $d = 40$ μm, and the output intensity I_B is equal to zero when no voltage is applied to the crystal. Assume that the applied voltage V is the sum of a bias voltage V_o plus a signal voltage V_1. Calculate: (i) the value of V_o required to bring the modulator in the linear regime; (ii) the upper limit to V_1, subject to the condition that the maximum output intensity does not exceed the limit of 75 % of the input intensity.

4.3 A light beam at $\lambda = 1550$ nm, linearly polarized along x, propagates along the optical axis (the z axis) of a *KDP* crystal. A voltage V is applied to the crystal along the z axis. After crossing the crystal the light beam goes through a sequence of a quarter-wave plate, having the optical axis in the x-z plane forming an angle of 45° with the x axis, and of a polarizer, with optical axis along y. Calculate the value of V required for a 100 % transmission of the incident beam through the sequence crystal-plate-polarizer.

4.4 An amplitude modulator based on the electro-optic Kerr effect can be made by putting the Kerr cell in between two crossed polarizers, as in Fig. 4.6. The external electric field E' is applied transversally, and the incident light is linearly polarized at 45° to the electric field. Calculate the half-wave voltage $V_{\pi/2}$ by assuming that the wavelength of light is 600 nm, the path-length of the light beam inside the Kerr cell is 5 cm, the distance between the two capacitor plates is 2 mm, and the electro-optic material is liquid nitrobenzene ($n_K^e - n_K^o = 2.6 \times 10^{-18}$ m^2/V^2).

4.5 A light beam at $\lambda = 1500$ nm goes through an acousto-optic cell made of flint glass ($us = 3$ km/s, $n = 1.95$) in which an acoustic wave of frequency 200 MHz is propagating. Calculate: (i) the wavelength difference between the diffracted wave and the incident wave; (ii) the deflection angle of the diffracted wave.

4.6 A light beam at $\lambda = 650$ nm goes through an acousto-optic cell made of water ($n = 1.33,\ p = 0.31,\ u_s = 1500$ m/s, $\rho = 1000$ kg/m^3) having a square cross-section. Calculate: (i) the frequency of the acoustic wave producing a deflection angle of 2°; (ii) the acoustic power required to transfer all the input power to the diffracted beam.

Chapter 5
Semiconductor Devices

Abstract The most important family of photonic devices is based on semiconductor materials. Semiconductor lasers have high efficiency and long lifetime, are fabricated by the same epitaxial growth techniques used in silicon microelectronics, can be designed to emit at almost any wavelength in the visible and near infrared range. In this chapter, after an introduction to the optical properties of semiconductors, the double-heterojunction structures used for semiconductor lasers are described, along with the main characteristics of laser emission. The subsequent sections deal with semiconductor amplifiers and light emitting diodes. The last part of the chapter concerns photodetectors and electro-absorption modulators.

5.1 Optical Properties of Semiconductors

The lasers described in Chap. 3 are based on transitions among energy levels of single atoms (or ions or molecules). In the case of semiconductor lasers the situation is different because the involved energy levels belong to the whole crystal, and not to a single atom.

In this section, a very simplified qualitative picture of the optical properties of semiconductors is given, with the aim of offering an intuitive description of the main processes involved in the interactions between a semiconductor crystal and an optical wave. From a phenomenological point of view semiconductors are solids whose electrical conductivity is typically intermediate between that of a metal and that of an insulator. Their conductivity can significantly increase by increasing the temperature or doping the material with impurities.

The atoms in a crystal lattice are strongly interacting. As a result, instead of individual energy levels, energy bands belonging to the whole crystal are formed. In the absence of external excitations, the bands are either fully occupied by electrons or are totally unoccupied (except for the special case of metals). The highest-energy fully occupied band is known as the valence band, while the lowest-energy unoccupied band is called the conduction band. These two bands are separated by a forbidden energy band, with bandgap energy E_g. Values of E_g for various semiconductors are given in Table 5.1.

V. Degiorgio and I. Cristiani, *Photonics*, Undergraduate Lecture Notes in Physics,
DOI 10.1007/978-3-319-20627-1_5

Table 5.1 Properties of some semiconductors

Semiconductor	E_g (eV)	λ_g (μm)	Type of gap
Ge	0.66	1.88	I
Si	1.11	1.15	I
AlP	2.45	0.52	I
AlAs	2.16	0.57	I
AlSb	1.58	0.75	I
GaP	2.26	0.55	I
GaAs	1.45	0.85	D
GaSb	0.73	1.70	D
GaN	3.45	0.36	D
InN	2.00	0.62	D
InP	1.35	0.92	D
InAs	0.36	3.50	D
InSb	0.17	7.30	D

In the fourth column I and D indicate, respectively, indirect bandgap and direct bandgap

At an absolute temperature of zero, an ideal semiconductor crystal has zero electric conductivity because there are no free charge carriers. An external energy source can excite an electron, causing a transition from the valence band to the conduction band. The transition leaves a hole in the valence band. Both electrons in the conduction band and holes in the valence band are mobile charge carriers contributing to the electric conductivity of the crystal.

The absorption of a photon with an energy of $h\nu \geq E_g$ excites an electron to the conduction band. By analogy with the case of an isolated atom, one could expect that the excited electron decays spontaneously to the valence band by emitting a photon. The situation inside the crystal is, however, more complicated, because the electron must be treated as a wave-like particle, with a wave-vector $\mathbf{k}$ and momentum $\mathbf{p} = h\mathbf{k}/(2\pi)$. Near the bottom of the conduction band, the energy versus momentum relation is approximately parabolic:

$$E = E_c + \frac{h^2k^2}{8\pi^2 m_c}, \tag{5.1}$$

where m_c, called effective mass of the electron, is different from the free electron mass m_e because of the effect of the interactions with the ions of the crystal lattice. A similar energy-momentum relation can be written for holes near the top of the valence band. The illustration in Fig. 5.1 qualitatively shows the band structures of two semiconductors, Si and $GaAs$. In the case of Si, the minimum of the conduction band and the maximum of the valence band do not occur at the same $\mathbf{k}$ value. A semiconductor that has such a band characteristic is called an indirect-bandgap semiconductor.

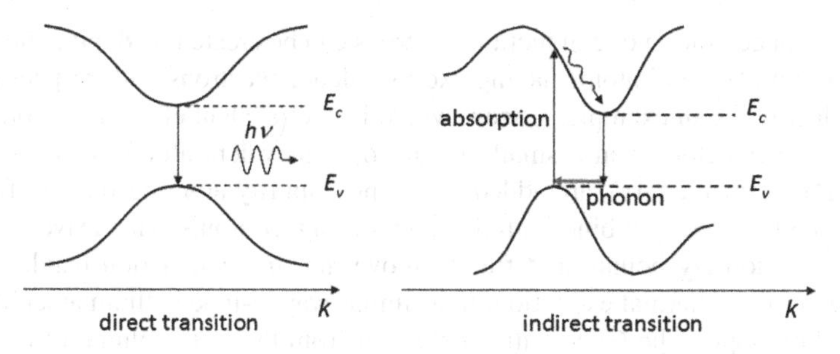

Fig. 5.1 Transitions in direct- and indirect-gap semiconductors

In contrast, a semiconductor like *GaAs* is a direct-bandgap semiconductor, because the minimum of the conduction band and the maximum of the valence band occur at the same **k** value. Table 5.1 indicates that both silicon (Si) and germanium (Ge) are indirect-bandgap materials, whereas several semiconductors made by combining a trivalent with a pentavalent atom, known as *III–V* semiconductors, have a direct bandgap. By absorbing a photon, the electron takes up not only the energy, but also the momentum of the photon. The order of magnitude of electron wave-vectors is $k \approx 2\pi/a$, where $a \approx 10^{-8}$ cm is the crystal lattice constant. The modulus of the photon wave-vector is $2\pi/\lambda$. In the case of visible or near infrared light, λ is about four orders of magnitude larger than a, so that it can be said that the electron momentum is practically unchanged by the photon absorption process. Therefore, as shown in Fig. 5.1, the transition from the valence band to the conduction band is vertical in the E, k plane.

As a general rule the excited electron will always go to occupy the lowest energy level in the conduction band. In correspondence, the hole will go to the highest available level in the valence band. To ensure momentum conservation, the decay of the excited electron to the valence band (also called electron-hole recombination) can be radiative (that is, a photon is emitted) only if the bottom of the conduction band corresponds to the top of the valence band, i.e. if they have the same wave-number. The maximum wavelength of light that can be absorbed (and emitted) by a given semiconductor is:

$$\lambda_g = \frac{hc}{E_g}. \tag{5.2}$$

Values of λ_g are reported in Table 5.1. In the case of indirect-bandgap, the excited electron quickly relaxes to the bottom of the conduction band, by making an intra-band transition in which energy and momentum are exchanged with the crystal lattice. At that point the electron-hole recombination is non-radiative, and involves the creation of a phonon.

In an undoped semiconductor (usually called an "intrinsic" semiconductor), at temperature T, there are always some electron-hole pairs, due to thermal excitation. The electron concentration n_i is equal to the hole concentration p_i. Both concentrations are small at room temperature, because $k_B T \ll E_g$.

A large concentration of conduction electrons can be created by doping the semiconductor with "donor" atoms having excess valence electrons with respect to the semiconductor. As an example, silicon, which is a tetravalent element, becomes an n-type semiconductor when a small amount of pentavalent atoms, such as phosphorus (P) or arsenic (As), are added to it. The impurity atom needs only four of its outer-shell electrons to bind with the neighboring Si atoms. This leaves its fifth electron only loosely bound, and free to move into the conduction band, after a room-temperature thermal excitation. In a similar way, n-type gallium arsenide can be created by doping the *GaAs* with an element from the sixth column of the periodic table, such as selenium (Se), which substitutes for an As atom in the lattice. To form p-type semiconductors with large hole populations, the crystal is doped with "acceptor" elements, having a deficiency of valence electrons with respect to the semiconductor.

By itself, a doped semiconductor is not better than the intrinsic one, from the point of view of light emission. In fact it can be shown that the product of the electron and hole concentrations, n and p, is constant for a given semiconductor at fixed temperature, that is, $np = n_i p_i$. The application of this law to an n-doped semiconductor shows that, when n is above n_i, p becomes smaller than p_i. Since the number of recombination processes is proportional to the product np, the number of emitted photons is independent of the doping of the semiconductor.

5.2 Semiconductor Lasers

Efficient light emission from a direct-bandgap semiconductor requires the presence of a large population of both type of carriers in the same crystal region. Such a condition can be realized in the depletion region of a p-n junction. Semiconductor lasers, which are currently the most important class of lasers, are based on forward-biased junctions. For this reason they are also known as diode lasers.

5.2.1 Homojunction Laser

In an n-type semiconductor at low temperature electrons occupy all the energy states of the valence band plus the lower energy states of the conduction band up to a maximum level, E_{Fn}, known as the Fermi energy. In a similar way, inside a p-type semiconductor electrons occupy the energy levels up to a maximum level, E_{Fp}, known as the Fermi energy of the p-type semiconductor, which is located inside the valence band. When a p-type and an n-type regions are brought into contact, electrons diffuse from the n-region into the p-region, and holes diffuse in the opposite direction. After a short transient period, in which recombination processes occur in a narrow region on both sides of the junction, a steady-state is reached when this region, known as the depletion-layer, is completely depleted of mobile charge carriers. A

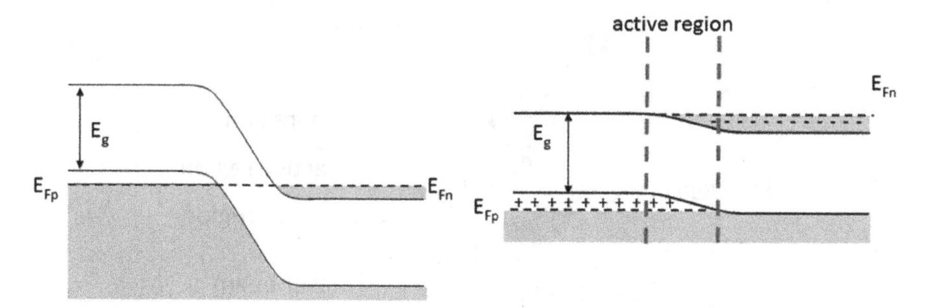

Fig. 5.2 Electron potential energy in an unpolarized (*left*) and directly polarized (*right*) *p*-*n* junction

double-layer of fixed charges, positive ions on the n-side and negative ions on the p-side, creates an electric field that blocks further diffusion of majority carriers. A positive potential difference between n-region and p-region is established, as shown in Fig. 5.2. At steady-state the two Fermi energies are equalized, $E_{Fn} = E_{Fp}$, in analogy to what happens in hydrostatics when a liquid fills to the same level in a set of communicating vases, regardless of their different shapes and volumes.

By applying a positive potential difference V between the p- and the n-regions, an electric current i flows across the junction. The current-voltage characteristic of the junction follows the law:

$$i = i_s \left[\exp\left(\frac{V}{V_o}\right) - 1 \right], \tag{5.3}$$

where i_s is the inverse saturation current, and $V_o = k_B T/e$. The inverse current is small, but non-zero, because minority carriers (electrons in the p region and holes in the n region) are always present at non-zero temperatures. The p-n junction acts as a diode. Diodes are unidirectional devices that are used in electronics for voltage rectification, logic circuits, and many other applications.

When the junction is forward-biased, that is $V \geq 0$, the flow of majority carriers across the junction produces large concentrations of both electrons and holes in the depletion region, causing a high number of recombination processes to occur. If the diode is fabricated from an indirect-bandgap semiconductor, like Si, then the energy from the non-radiative recombination is transferred to the crystal lattice as vibrational energy. If, however, the junction is made with a direct-bandgap material, then the recombination is radiative, and the depletion region becomes a light emitter in a wavelength range close to λ_g. At low current density, mainly spontaneous recombination processes occur. However, at large enough carrier concentrations, stimulated electron-hole recombination processes become important, making optical amplification possible.

As shown in Fig. 5.2, the forward bias makes the two Fermi levels unequal, E_{Fn} becomes larger than E_{Fp}. The difference $\Delta E_F(J) = E_{Fn} - E_{Fp}$ grows as a function of the current density J flowing across the junction. It is possible to have optical gain

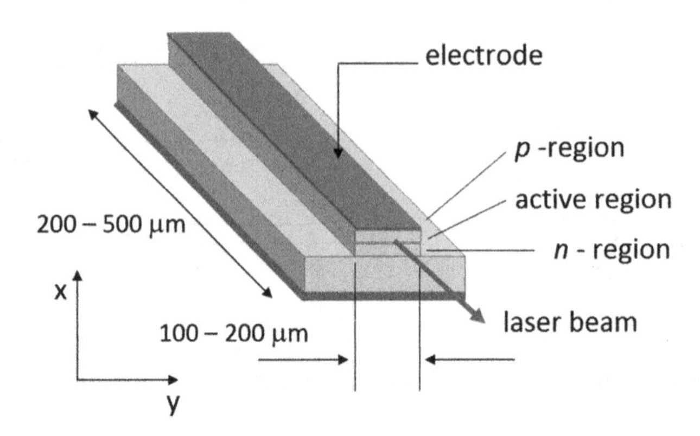

Fig. 5.3 Structure of the homojunction laser

only if $\Delta E_F(J)$ becomes larger than E_g. Precisely, light emission occurs in the frequency band determined by the condition:

$$\Delta E_F(J) \geq h\nu \geq E_g. \tag{5.4}$$

Equation (5.4) can be seen as the condition for population inversion in the current-injected semiconductor.

The first semiconductor laser was operated in 1962 using a *GaAs* *p-n* junction. Since a junction in which the same material is used for both the p and the n sides is called a homojunction, this laser is known as a homojunction laser. The light emission was in the near infrared, at a wavelength close to $\lambda_g = 0.85\ \mu$m. The basic structure of the homojunction laser is shown in Fig. 5.3. The electric current flows along the x axis. The active region is the depletion layer with a length of 0.2–0.5 mm and a width, in the y direction, of 0.1–0.2 mm. The thickness of the active layer along x is very small, typically around 1 μm.

The feedback needed for laser action inside the Fabry-Perot cavity is provided by the reflections from the two parallel end faces, which are prepared by cleavage along a crystal plane. Since the refractive index of a semiconductor is fairly large, the refractive-index mismatch at the air-semiconductor interface gives a sufficiently high reflectance without the need for reflective coatings of the end facets. As an example, the refractive index of *GaAs* is $n = 3.6$, so that the reflectance from (2.34) is $R \approx 30\,\%$.

The homojunction laser was operated continuously at the liquid nitrogen temperature ($T = 77$ K), with a threshold current density $J_{th} \approx 1$ kA/cm^2. When T is increased, more and more electrons can occupy energy states above Fermi levels. This has the effect of reducing the optical gain. It can be shown that J_{th} grows exponentially with T, according to the law:

$$J_{th}(T) = J_{th}(T_o) \exp\left(\frac{T}{T_o} - 1\right). \tag{5.5}$$

At room temperature, J_{th} becomes of the order of 100 kA/cm^2. For such large current densities, the junction is very quickly damaged, so that only pulsed operation is possible. A drawback of the homojunction laser is that the size of the laser beam in the x direction is much larger than the thickness of the active region. As a consequence, a large fraction of the beam power travels inside the p and n regions, where it is absorbed. Another drawback is that the charge carriers injected inside the depletion region quickly drift outside this region, so that only a small fraction is available for stimulated electron-hole recombination.

5.2.2 Double-Heterojunction Structures

The performance of the semiconductor laser was greatly improved by using a double heterojunction, instead of the homojunction described in the preceding section. The double-heterojunction laser has an active medium sandwiched between p and n materials that differ from the active material. This laser was invented around 1970 by Alferov and Kroemer, who were awarded the Nobel prize for physics in the year 2000. As shown in Fig. 5.4, the active region *p-GaAs* is sandwiched between the regions $p\text{-}Ga_{1-x}Al_xAs$ and $n\text{-}Ga_{1-x}Al_xAs$, where x is the fraction of aluminum atoms. Typically, $x = 0.3$. Such a structure has many positive features, as illustrated in Fig. 5.5. The bandgap of the active layer ($E_{g1} = 1.5$ eV for *GaAs*) is smaller than that of the cladding layers ($E_{g2} = 1.8$ eV for $Ga_{0.7}Al_{0.3}As$). Therefore, the "wings" of the laser beam are not absorbed by the cladding regions. As a general rule for *III–V* semiconductors, a change in composition that leads to an increase in bandgap also results in a reduced refractive index. In fact, the refractive index of $Ga_{0.7}Al_{0.3}As$ is 3.4, smaller than that of *GaAs*, so that an optical waveguide effect exists for the laser beam. An additional benefit of the double-heterojunction is that the difference between the bandgaps of the two junctions creates an energy barrier that increases the carrier concentration in the active region. By increasing the carrier concentration, the optical gain increases.

By using the double-heterojunction structure, the room-temperature threshold current density is reduced by two orders of magnitude with respect to the homojunction structure, reaching a value $J_{th} \approx 0.5$ kA/cm^2.

In order to form a double-heterostructure without introducing material strain, a very important condition must be satisfied, namely that the lattice period of the active layer must closely match that of the cladding layers. A mismatch gives rise to the

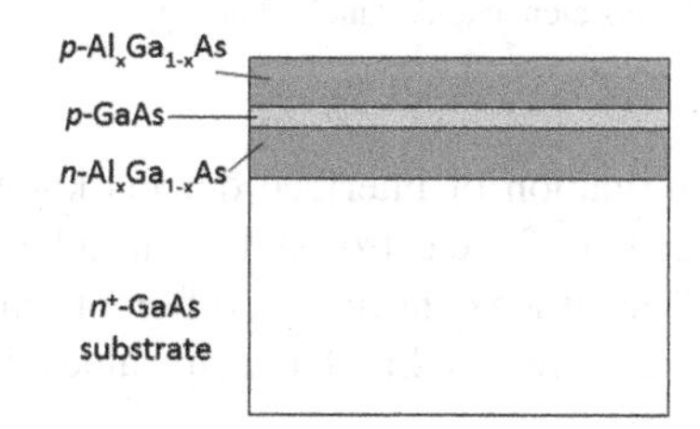

Fig. 5.4 Double-heterojunction in *GaAs*

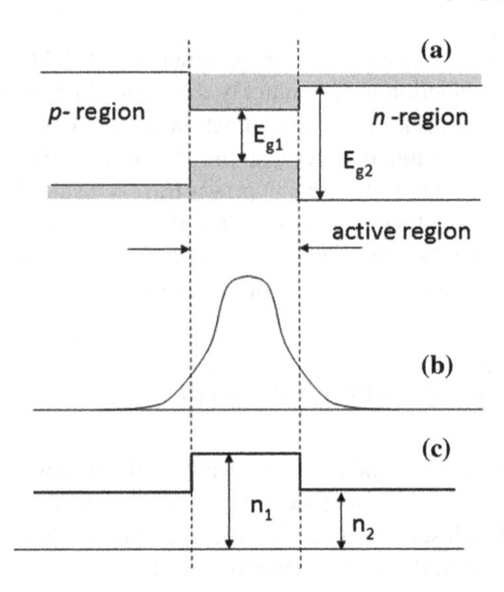

Fig. 5.5 Double heterojunction: **a** behavior of the energy bands; **b** transverse profile of the laser beam; **c** refractive index profile

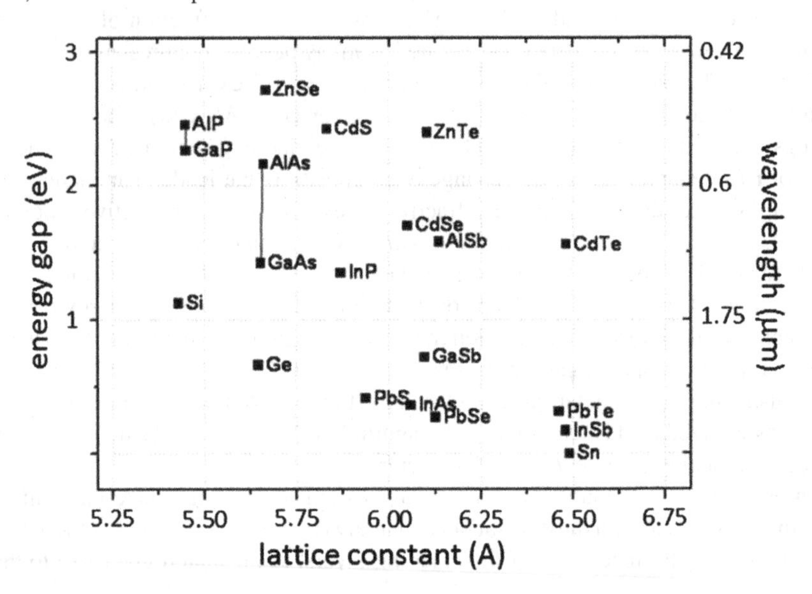

Fig. 5.6 Lattice constants of different semiconductors. Taken from Opensource Handbook of Nanoscience and Nanotechnology

formation of interface defects known as misfit dislocations. These defects act as centers for electron-hole non-radiative recombination that reduce the optical gain. The lattice constants of different semiconductors are given in Fig. 5.6, together with the corresponding bandgap values. Since the radii of *Ga* and *Al* ions are almost the

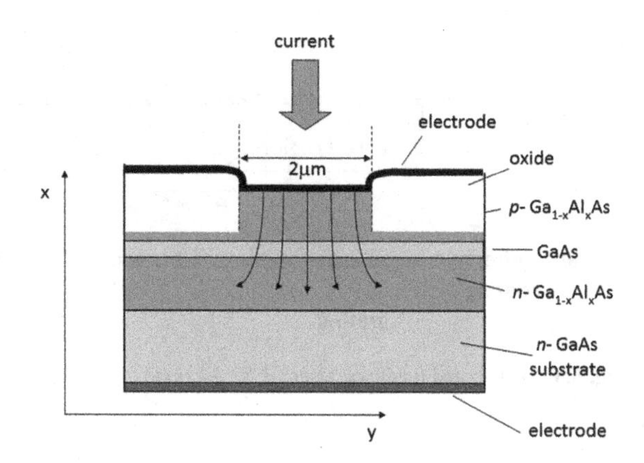

Fig. 5.7 The double-heterojunction laser: stripe-geometry configuration

same, the lattice periods of *GaAs* and *AlAs* coincide, and thus lattice-matching for the *GaAs*/*GaAlAs* structure is very good. In the case of the quaternary compound $Ga_{1-x}In_xAs_{1-y}P_y$, which is used for optical-communications lasers, the alloy can be lattice-matched to cladding layers made of *InP* by choosing a specific y/x ratio. It is found that the optimal value is $y = 2.2\,x$.

Stripe geometry. Double-heterojunction lasers often use the so-called stripe-geometry configuration shown in Fig. 5.7. By introducing an insulating oxide layer, current from the positive electrode is constrained to flow in a narrow stripe. By reducing the cross-section, the total current is correspondingly reduced for a given current density. Another method to reduce the cross-section is that of creating a refractive index profile that produces a confinement along the y-direction, as is the case for both the "channeled substrate laser", and "buried heterostructure laser" structures. The size of the active region along y can be as small as a few μm, with a threshold current, i_{th}, around 10 mA and applied voltage of 1.5–2 V.

Distributed feedback. As it has been discussed in Sect. 3.6, a Fabry-Perot laser can operate with more than one longitudinal mode. The actual number of oscillating modes is determined by the ratio between the frequency band $\Delta\nu_L$ of the amplifier and the distance $\Delta\nu$ between two adjacent longitudinal modes. In several applications, among which optical communications is the most important, emission at a single frequency is necessary. Among the different methods to realize a single-mode semiconductor laser, the distributed feedback geometry is the most widely used. Figure 5.8 shows the scheme of a distributed feedback laser, also known as DFB laser, which consists of an active medium in which a periodic thickness variation is introduced in one of the cladding layers. The periodic corrugation provides optical feedback for the laser oscillation, and the end facets are coated with antireflection coatings. The laser behaves as an optical amplifier in which the forward and the backward beam have a frequency-selective coupling. These two counter-propagating beams are efficiently coupled to each other only if their wavelength is such that

$$\lambda = 2n_{eff}\Lambda \tag{5.6}$$

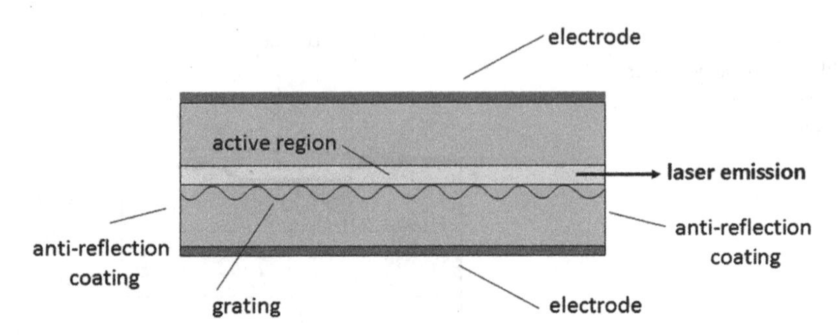

Fig. 5.8 Distributed feedback laser

where Λ is the pitch of the periodic corrugation and n_{eff} is an effective refractive index. The exact value of n_{eff} depends on specific details of the structure. The fabrication of a high-quality uniform grating is rather difficult, because the pitch has submicron periodicity. For example, in the case of a 1550-nm *InGaAsP* laser with $n_{eff} = 3.4$, one finds that $\Lambda = 0.23\ \mu\text{m}$.

Surface-emitting structure. The area of semiconductor lasers is in continuous evolution, and many different types of structures have been tested. The semiconductor lasers discussed so far are edge-emitting, i.e. they generate an optical beam propagating parallel to the junction plane. However, in some cases, it is interesting to have laser emission normal to the junction plane. Such lasers are known as vertical-cavity surface-emitting lasers, or VCSELs. Compared to the edge-emitting laser, the VCSEL has a much shorter length of active medium, and a much broader emitting area. Since the single-pass gain is low, high-reflectivity mirrors must be used to avoid high threshold currents. As shown in Fig. 5.9, such mirrors are made by alternating several layers of two different semiconductors, following the multilayer scheme described in Sect. 2.2.4. Due to the short length of the laser cavity, the spac-

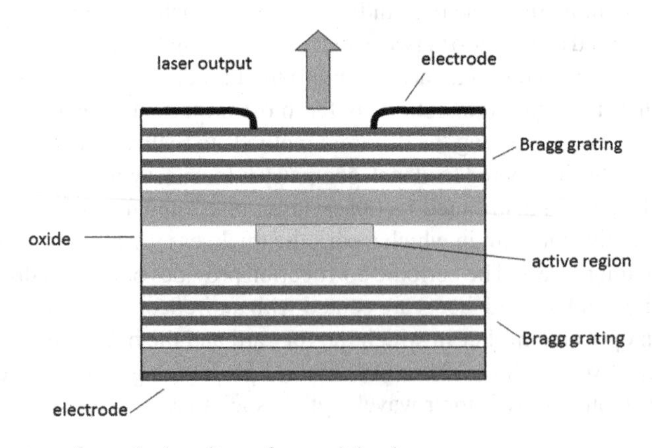

Fig. 5.9 Structure of a vertical-cavity surface-emitting laser

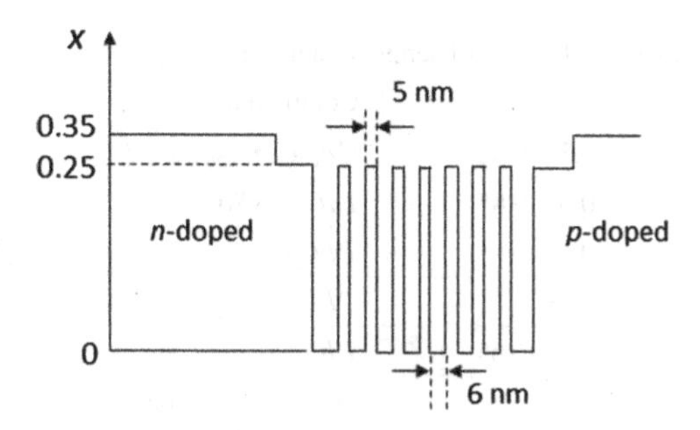

Fig. 5.10 Active layer of a semiconductor laser made of a multiple quantum well, with a composition given by $In_{0.5}Ga_{0.5}P/In_{0.5}Ga_{0.5-x}Al_xP$. The plot represents the behavior of the Al fraction, x, across the junction

ing of $\Delta\nu$ between adjacent longitudinal modes is very large. Thus, if the frequency of one mode coincides with the peak reflectivity of the multilayer mirrors, the two adjacent modes fall outside the high-reflectivity band of the mirrors, and single-longitudinal-mode operation is achieved. By limiting the diameter of the emission area to 5–10 μm, a very low threshold current is obtained ($i_{th} \approx 1$ mA). Correspondingly, the output power is also small, typically of the order of 1 mW. However, it is possible to build large-area (300 μm diameter) VCSELs with output power of watts.

Multiple-quantum-well. A very interesting concept recently emerged in many distinct scientific and technological areas is the use of artificially tailored structures at the nanoscale. It is beyond the scope of this text to explain the properties of nanostructured materials, but the important aspect is that electrons confined in nanostructures have discrete energy levels, instead of continuous, with the energy of those levels fixed by the geometry of the structure, following the rules of quantum mechanics. As an example, Fig. 5.10 shows the nanostructured active layer of a semiconductor laser, in which the active region consists of a number of alternating layers of narrow and wide bandgap materials. The thickness of the narrow-bandgap layers (the "quantum wells") and that of the wide-bandgap barriers are both about 5 nm. A laser of this type is known as a "multiple quantum well" (MQW) laser. MQW lasers are realized in both edge-emitting and surface-emitting configurations. In general, they have a very low threshold current density and a high efficiency. They also have a very stable emission wavelength, not influenced by changes in electric current or temperature.

5.2.3 Emission Properties

Laser emission has been obtained from many different *III–V* semiconductors. Table 5.2 lists the most common semiconductor lasers. The trivalent atom can be gallium or aluminum or indium. The pentavalent atom can be arsenic or phosphorus or antimony (Sb) or nitrogen (N). By using combinations of two trivalent atoms and/or two pentavalent ones, the emission wavelength can be tuned, in some cases,

Table 5.2 Operating wavelength range of semiconductor lasers

Compound	λ (μm)	Compound	λ (μm)
$In_xGa_{1-x}N$	0.37–0.50	$Ga_{0.42}In_{0.58}As_{0.9}P_{0.1}$	1.55
$Ga_{1-x}Al_xAs$	0.62–0.85	$InGaAsSb$	1.7–4.4
$Ga_{0.5}In_{0.5}P$	0.67	$PbCdS$	2–4
$Ga_{0.8}In_{0.2}As$	0.98	$PbSSe$	4.2–8
$Ga_{1-x}In_xAs_{1-y}P_y$	1.10–1.65	$PbSnSe$	8–32
$Ga_{0.27}In_{0.73}As_{0.58}P_{0.42}$	1.31	Quantum cascade	2–70

over a very broad range. This is a distinctive property of semiconductor lasers: solid-state and gas lasers operate at discrete wavelengths fixed by the position of atomic or ionic energy levels (with a few exceptions), but semiconductor lasers can be designed to operate at essentially any wavelength in the visible and near infrared.

The wavelength range 0.62–0.85 μm is covered by devices using $Ga_{1-x}Al_xAs$. Since the lattice constant of $Ga_{1-x}Al_xAs$ does not depend on the fraction x of Al atoms (see Fig. 5.6), heterojunctions containing any fraction of Al can be grown epitaxially. A limit to using large values of x comes from the fact that the compound acquires an indirect bandgap when $x \geq 0.7$. $Ga_{1-x}Al_xAs$ lasers emitting at 0.81 μm have an important application for pumping *Nd-YAG* lasers, as described in Sect. 3.9. Optical-communication lasers, emitting around 1.31 or 1.55 μm are based on the quaternary compound $Ga_{1-x}In_xAs_{1-y}P_y$ grown on top of *InP*. By changing the fractions x and y, it would be possible to cover a wide range of bandgap energies, from 0.4 to 2.2 eV. In practice, because of the mismatch in atomic sizes between gallium and indium, and also between arsenic and phosphorus, only the range 0.7–1.4 eV ($\lambda = 0.88$–1.76 μm) can be usefully covered. The $Ga_{0.8}In_{0.2}As$ laser, emitting at 0.98 μm, has an important role as the pump of the fiber-optic amplifier described in the following chapter. *III-V* semiconductors using nitrogen as the pentavalent atom exhibit a large bandgap, and so can emit blue and violet light. In particular, the wavelength range 0.37–0.50 μm is covered by the $In_xGa_{1-x}N$ laser, which is useful for a variety of applications, such as writing and reading Blu-ray optical discs.

By using *II–VI* semiconductors, lasers operating in the mid- and far-infrared range of wavelengths can be made. Sources in this range of wavelengths are useful for such applications as chemical sensing and trace-gas analysis. Since the laser efficiency strongly decreases when the photon energy becomes comparable to k_BT, all these devices work at cryogenic temperatures. The same wavelength range can be covered much more efficiently by a new family of semiconductor lasers, termed quantum cascade lasers (QCL), which make use of MQW structures in *III–V* semiconductors. It goes beyond the scope of this text to describe the physical mechanism of QCLs, which is completely different from that of diode lasers. It suffices to mention that the QCL can be designed to emit at any wavelength in the mid- and far-infrared, through appropriate choice of the nanostructure. The QCL can operate at room temperature as a continuous source or as a source of picosecond pulses.

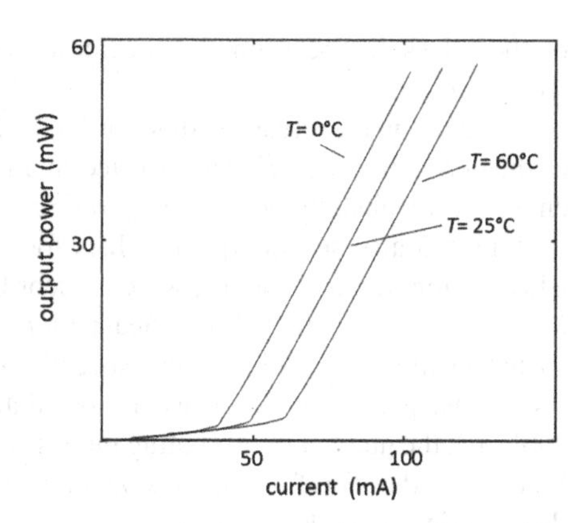

Fig. 5.11 Output power versus input current for a double-heterojunction laser, at three different temperatures

A double-heterojunction laser has an internal efficiency, defined as the ratio between the number of radiative recombinations and the number of injected electron-hole pairs, that can be close to 100 %. The overall efficiency of the laser, defined as the ratio between emitted optical power and injected electric power, often termed "wall-plug efficiency", can be as large as 50 %.

Experimental curves of output power versus input current, at three different temperatures, are shown in Fig. 5.11. It is found that the threshold current grows with increasing the temperature, as predicted by (5.5) with a best-fit parameter $T_o \approx 160$ K.

The output beam of a semiconductor laser is usually strongly diverging, because of the smallness of the transverse size of the active region. Typically, the beam observed at the output facet of a *GaAs* laser has an elliptical cross-section with a size of 1 μm in the direction orthogonal to the junction (x direction), and of 10 μm in the y direction. The far-field angular spread due to diffraction is about 50° along x and 5° along y. By using a cylindrical lens with two different focal lengths along x and y, the beam can be collimated to obtain a circular spot having a divergence angle of just a few milliradians.

The spectrum of the laser emission has a peak wavelength λ_p that depends on pump current and temperature. The wavelength interval $\Delta\lambda_L$ over which laser oscillation can occur grows as the pump current increases, as described by (5.4). Some typical values of $\Delta\lambda_L$ are reported in Table 5.3. If $\Delta\lambda_L$ grows and, at the same time, the highest emission wavelength remains fixed to hc/E_g, then the central wavelength

Table 5.3 Spectral properties of three semiconductor lasers

	$\lambda = 850$ nm	$\lambda = 1300$ nm	$\lambda = 1550$ nm
$\Delta\lambda_L$ in nm	2.0	5.0	7.0
$\Delta\lambda$ in nm	0.27	0.6	0.9
$d\lambda_p/dT$ in nm/°C	0.22	0.5	0.73
$d\lambda_q/dT$ in nm/°C	0.06	0.12	0.18

of the emission spectrum, λ_p, decreases with increasing pump current, at fixed temperature.

Since E_g is a decreasing function of T, λ_p grows with T. Some typical values of the derivative $d\lambda_p/dT$ are reported in Table 5.3. Clearly, in order to stabilize the emission wavelength, precise temperature control is needed.

Table 5.3 also shows typical values for the wavelength separation between two adjacent longitudinal modes, which, according to Sect. 3.6, is expressed as $\Delta\lambda = \lambda_q - \lambda_{q-1} \approx {\lambda_p}^2/(2L')$. The optical path L' is very short in the semiconductor laser, so that the modes are rather widely spaced. However the bandwidth $\Delta\lambda_L$ is also large. As a consequence, the laser usually oscillates on several modes. Using the data in Table 5.3, the number of oscillating modes, as calculated from $N = \Delta\lambda_L/\Delta\lambda$, is 7–8. Since the index of refraction of semiconductors increases with T, the wavelength of the generic q_{th} mode also increases with T. Values of $d\lambda_q/dT$ are reported in the last row of the same table.

The CW output power of a semiconductor laser can only be several hundreds of milliwatts, at most. By fabricating a two-dimensional array of lasers on a single substrate, and then coupling all the outputs into a single optical fiber, a monochromatic optical power of several tens of watts can be achieved. This scheme is often used to pump solid-state and fiber lasers.

An important feature of a semiconductor laser is its very long lifetime: as an example, a test conducted on a set of *GaAlAs* lasers, operating at 70 °C with an output power of 5 mW, has measured a lifetime of 30,000 h (3.5 years of continuous operation). *GaInAsP* lasers have been found to have even longer lifetimes.

By pulsing the pump, a semiconductor laser can be made to emit short pulses, for use, as an example, in portable optical radars. It can also be intensity modulated by controlling the pump current. The modulating signal (typically 40–60 mA peak-to-peak) is superposed to the direct bias current. The maximum frequency of modulation depends on how quickly the carrier density in the active region can be varied. Typical values are of the order of a few GHz, but, in specially designed devices, modulation frequencies of 20–30 GHz have been demonstrated.

Diode lasers can also generate ultrashort pulses, using either active or passive mode-locking. Typically, pulse-widths of between 0.5 and 5 ps with repetition rates of between 1 and 100 GHz can be obtained. For repetition rates below roughly 10 GHz, an external cavity setup is generally required. Particularly interesting are the vertical-external-cavity surface-emitting lasers (VECSELs), which have a vertical-cavity surface-emitting structure with one of the two high-reflecting mirrors external to the diode structure. Typical cavity lengths are from a few millimeters up to tens of centimeters. The gain chip generally contains a highly reflective bottom section to reflect the laser and pump light, an active semiconductor gain section, and an anti-reflective top layer. By passively mode-locking optically pumped VECSELs, it is possible to obtain much higher pulse energies than those generated by edge-emitting structures.

5.3 Semiconductor Amplifiers

A semiconductor optical amplifier, usually known as SOA, is realized by eliminating the feedback elements in a semiconductor laser, usually by putting an anti-reflection coating on the two facets of the active region. Of course, the operating wavelengths of SOAs are the same as those of the corresponding lasers, listed in Table 5.2.

Typically, at $\lambda = 1.55$ μm, the gain per unit length g is of the order of 10^5 m^{-1} and the gain bandwidth is about 10 THz. Since g is somewhat dependent on the polarization state of the input signal, a simple SOA does not generally preserve the input polarization. However, it is possible to implement more complex amplifying schemes that are polarization-preserving.

SOAs are attractive for many applications because they are small, have a large optical gain, and can be driven by a low-voltage supply. However, the small size of the active region makes the input coupling rather critical, while the large gain can give rise to signal distortions because of saturation effects.

The most important applications of SOAs are in the optical-communication area, as will be discussed in Chap. 8.

5.4 Electroluminescent Diodes

The electroluminescent diodes fabricated from semiconductors are known as light emitting diodes (LEDs). An LED is a solid-state light source using the spontaneous emission generated by radiative electron-hole recombinations in the active region of the diode under forward bias. Unlike a laser, an LED does not have an optical cavity or a threshold current. It starts emitting light as soon as a forward bias voltage is applied to the junction, and emits incoherent and unpolarized spontaneous photons. As in a laser, the emission wavelength is determined by the bandgap energy of the used semiconductor. An other important electroluminescent diode is the organic light-emitting diode, known as OLED, which is based on organic compounds.

5.4.1 LEDs

The first visible LED, emitting at 710 nm, was demonstrated by Holonyak in 1962 using a *GaAsP* diode. After this realization the research concerning electroluminescent materials was extended to many different *III–V* semiconductors. An important step was made in 1994 with the invention of the blue LED. Akasaki, Amano, and Nakamura were awarded the Nobel prize in Physics in the year 2014 for this invention. The light emitted by an LED has a wavelength $\lambda_g = hc/E_g$, and a bandwidth $\Delta\lambda$ given by:

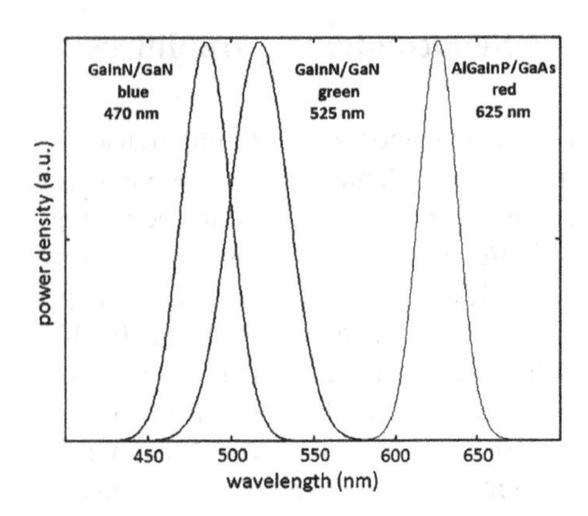

Fig. 5.12 LED emission in the three fundamental colors

$$\Delta\lambda = \frac{2k_B T {\lambda_g}^2}{hc} \tag{5.7}$$

As an example, an LED emitting 500-nm light at room temperature, has an emission bandwidth predicted by (5.7) of about 10 nm. Commercially available LEDs cover the spectral range from the near ultraviolet to the near infrared, with wavelengths ranging from about 370 nm to 1.65 μm. As shown in Fig. 5.12, red-orange LEDs use $GaAs_x P_{1-x}$, and blue-green LEDs use $In_{1-x}Ga_x N$. The emission wavelength is tuned by changing the composition, which modifies the parameter of x. Some LEDs even use indirect-bandgap semiconductors doped with impurities although these offer lower efficiency.

The structure of a surface-emitting LED is shown in Fig. 5.13. Spontaneous emission radiates in all directions, but the downward-emitted light is absorbed, and the extraction of the upward-emitted light is possible only over a limited solid angle. The limitation to extraction efficiency is due to reflection from the semiconductor-air interface. The spontaneous photons that reach this interface at an angle of incidence larger than the limit angle θ_c are totally internally reflected. If the refractive index of the semiconductor is 3.4, then this critical angle is just 17°, and so the extraction efficiency is very low. Various techniques can be used to increase the extraction efficiency: the reflectance of the semiconductor-air interface can be reduced by an anti-reflection coating, and the downward-emitted light can be recuperated by using a transparent substrate associated to a highly-reflecting bottom surface.

An important parameter that characterizes a light source is the ratio between emitted optical power and injected current. For a visible LED, this ratio is around 0.5 W/A. The overall LED efficiency, η, can be as large as 25 %. As a comparison, the efficiency of an incandescent lamp is about 1 %. High efficiency also implies less heat generation and less damage. The lifetime of an LED can reach 50,000 h of continuous operation.

LEDs are currently used in a large variety of applications, some of which are described in Chap. 8. When an LED replaces traditional lamps, then white light emission is usually required. White light can be obtained by mixing the emissions

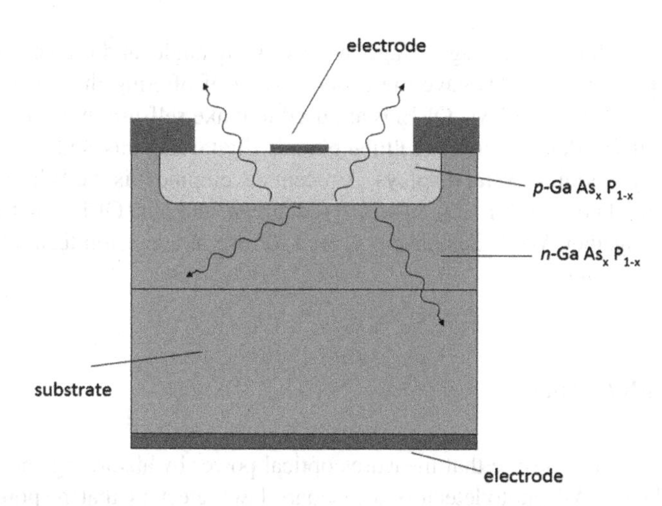

Fig. 5.13 Structure of a *GaAsP* LED. A layer of Te-doped $n{-}GaAs_{0.6}P_{0.4}$ is grown on the substrate $n{-}GaAs$. The double heterojunction is formed by superposing a layer of Zn-doped $p{-}GaAs_{0.6}P_{0.4}$. The LED emits in the *upward* direction, while the *downward* emitted light is absorbed by the substrate

of three LEDs operating at the three fundamental colors (red-green-blue, RGB). However, in practice, it is often more convenient to make use of a single blue-emitting LED coated with phosphor molecules. The phosphors absorb the blue emission, and spontaneously re-emit light of a longer wavelength. By mixing different phosphors with different colors, white light emission is achieved. The design and fabrication of this white light source is both simpler and cheaper than a complex RGB system.

5.4.2 OLEDs

In an organic light-emitting diode (OLED), the electroluminescent material is an organic compound, which is electrically conductive and behaves as an organic semiconductor. Two thin films of organic semiconductor, forming an organic heterostructure, are sandwiched between two electrodes on a transparent substrate. Electronic charges are transported and injected into the heterostructure from the electrodes: electrons from the cathode, and holes from the anode. In analogy to the case of semiconductor devices, the radiative recombination of electron and hole generates light. The wavelength of this emission depends on the band gap of the organic material. OLEDs can be used to produce light of a very wide range of wavelengths by modifying the precise structure of the organic material. OLEDs are fabricated from small organic molecules or conjugated polymer chains, by using standard techniques of thin film deposition. These are simpler and less expensive than the epitaxial growth used for fabricating LEDs. The potential applications of OLEDs include thin, low-cost displays with a low driving voltage, wide viewing angle, and high contrast and color gamut. Polymer LEDs have the added benefit of offering the possibility for printable and flexible displays. OLEDs are used to make self-luminous displays for portable electronic devices such as cellular phones, digital cameras, and as a source of backlighting for liquid-crystal displays. A recent development is the White Organic Light-Emitting Diode (WOLED), in which red, green, and blue OLED materials are sandwiched together. WOLEDs have potential for use in television technology and architectural lighting.

5.5 Photodetectors

A photodetector is a device that measures optical power by absorbing energy from an optical beam. All photodetectors are square-law detectors that respond to the intensity, rather than the field amplitude, of an optical signal. Commonly used photodetectors are based on either thermal effects or electric effects.

Thermal detectors convert the absorbed energy into heat. In a bolometer, light is absorbed by a metallic or semiconducting resistor, or by a termistor. The temperature variation induces a change in the electric resistance of the absorbing material, which is measured, for instance, by a Wheatstone bridge. Highly sensitive modern bolometers use superconducting absorbers operating at cryogenic temperatures. Another possibility is to use a thermocouple, which is typically made by a ring with two bimetallic junctions. If one junction is kept at fixed temperature, and the other is heated by light absorption, then an electric current flows inside the ring, because of an effect known as thermoelectric effect or Seebeck effect. The intensity of the electric current is proportional to the absorbed optical power. Thermocouples are self powered and require no external form of excitation. In general, the response of thermal detectors does not depend on the wavelength of the incident radiation. However, they have a slower response time and lower sensitivity, compared to electric detectors.

More important then thermal detectors are the detectors that use electric effects, which include photoelectric detectors, photoconductors, and photodiodes. They can be made to be extremely sensitive, and can have response times capable of following very fast optical signals.

5.5.1 Photoelectric Detectors

The photoelectric effect is the emission of an electron when a material absorbs energy from a light beam. The emitted electrons are called photoelectrons. The effect was discovered by Hertz in 1887 and explained by Einstein in 1905, for which he was awarded the Nobel Prize in 1921. The photoelectron is generated only if the photon energy $h\nu$ satisfies the condition:

$$h\nu \geq E_i, \tag{5.8}$$

where E_i is the ionization energy of the electron inside the material. In the case of solids, E_i is usually known as the "work function". The excess energy, $h\nu - E_i$, is transformed in kinetic energy of the photoelectron. The material can be a gas, a liquid, or a solid. Considering isolated atoms in the ground state, low ionization energies are found for elements belonging in the first column of the periodic table, i.e. the alkali metals. This is not surprising because these elements have only one electron in their outer electron shell. The cesium atom has the lowest binding energy of $E_i = 3.8$ eV. The highest binding energy is 24.5 eV for helium, a noble gas. Equation (5.8) states that a photoelectron can only be generated if the wavelength of the photon is below a threshold value given by:

$$\lambda_{th} = \frac{hc}{E_i}. \tag{5.9}$$

For the cesium atom, $\lambda_{th} = 327$ nm, which is in the ultraviolet. While no isolated atom emits photoelectrons under excitation from visible light, the situation is more favorable for solids. The work function of solid cesium is 1.94 eV, corresponding to a threshold wavelength of $\lambda_{th} = 640$ nm. Even lower values of the work function are found for cesium compounds, such as cesium antimonide. The threshold wavelength can reach a value of about 1.1 μm, extending the applicability of photoelectric detectors to the near infrared.

Photoelectric cell. A photoelectric cell, also called a vacuum photodiode, can be made by collecting the photoelectrons emitted by the photosensitive surface of a negative electrode (photocathode) at a positive electrode (anode). The quantum efficiency of the photosensitive surface layer, η, defined as the ratio between the number of photoelectrons and the number of absorbed photons, can vary between 10^{-4} and 10^{-1}, depending on the incident wavelength and the nature of the surface layer. The two electrodes are placed in a vacuum tube with a transparent window. An electric current flows in a closed circuit configuration under illumination of the photocathode. The photocurrent i_{ph} is obtained by multiplying the electron charge by the number of photoelectrons collected by the anode per second:

$$i_{ph} = \frac{\eta e P}{h\nu}, \tag{5.10}$$

where P is the incident optical power. The sensitivity of the photoelectric cell is expressed by the ratio i_{ph}/P (Ampère/Watt). As an example, if $h\nu = 2$ eV (red light) and $\eta = 0.1$, then the sensitivity is 50 mA/W.

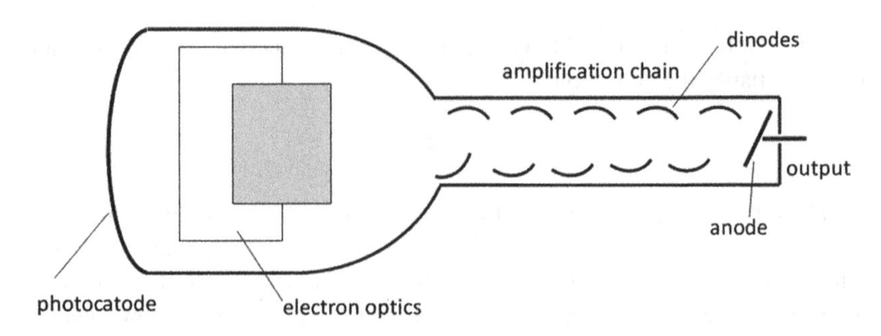

Fig. 5.14 Structure of a photomultiplier tube

Photomultiplier tube. The sensitivity of a photoelectric detector can be improved by many orders of magnitude by multiplying the number of emitted electrons via the process of secondary emission. This occurs when the accelerated electrons impact intermediate electrodes, called dynodes, maintained at ever-increasing potential-differences. The resulting device, known as a photomultiplier tube, is schematically illustrated in Fig. 5.14. The photocathode contains combinations of materials such as cesium, rubidium and antimony, specially selected to provide a low ionization energy. The number of dynodes varies between 10 and 14. Their shape and position are designed in order to guide the electrons along trajectories which maximize the collection efficiency. Typically the potential difference between two successive dynodes is around 100 V and the supply current is around 1 mA. A photoelectron emitted by the photocathode generates m secondary electrons when it impacts the first dynode. Each secondary electron, when collected by the second dynode, further produces secondary electrons. If there are n dynodes, then the number of electrons arriving at the anode is m^n. Assuming $m = 4$ and $n = 10$, the gain of a photomultiplier tube is: $G = m^n = 4^{10} \approx 10^6$. With 14 dynodes, the gain can approach 10^8. If the quantum efficiency is the same, then the sensitivity of a photomultiplier tube is larger than that of a photoelectric cell by a factor G.

The photoelectric effect has a very short intrinsic response time, comparable to the oscillation period of the optical field. The response time of the device is limited by two effects: (i) there is a spread in the transit time of the electrons through the multiplication structure, such that the G electrons produced by a single photoelectron do not simultaneously reach the anode, but arrive with a spread of a few nanoseconds; (ii) the device has an equivalent capacitance, C, between the anode and the ground that gives a time constant of $R_L C$, because of the non-zero load resistance R_L. If $C = 20$ pF and $R_L = 50\ \Omega$, then the time constant is 1 ns.

A photomultiplier tube with a gain $G = 10^8$ produces an easily measurable output pulse, in response to the absorption of a single photon. Indeed G electrons bunched in a time interval τ give a current pulse with a peak value Ge/τ. Therefore the voltage pulse measured across the load resistance has a peak value GeR_L/τ. If $G = 10^8$, $\tau = 1$ ns, $R_L = 50\ \Omega$, then the peak voltage is 0.8 V. When observing the single-photoelectron pulses, very large amplitude variations are found from one pulse to the other, because m is not a fixed number, but a statistical variable with a Poisson probability distribution.

When the optical power incident on the photocathode is so low that the number of photoelectrons per unit time is less than τ^{-1}, the electric signal at the photomultiplier output appears as a discrete sequence of pulses. The experimental observation of this discreteness has a great conceptual relevance, because it demonstrates that electromagnetic energy is absorbed by matter in quantum steps.

5.5.2 Semiconductor Photodetectors

Semiconductor photodetectors are based on the generation of an electron-hole pair caused by the absorption of a photon. Sometimes this effect is called "internal photoelectric effect", whereas the electron extraction process described in the preceding section is called external photoelectric effect. The photogeneration of electron-hole pairs can be exploited for photodetection purposes in different ways.

Photoconductors. The conductivity of a semiconductor that has electron and hole concentrations of n and p, respectively, is:

$$\sigma = e(\mu_e n + \mu_h p), \tag{5.11}$$

where μ_e and μ_h are the electron and hole mobilities, respectively. When a semiconductor is illuminated with light of a sufficient photon energy, carriers in excess of the equilibrium concentrations are generated. As a consequence, the conductivity increases by an amount that is proportional to the number of photogenerated electron-hole pairs. Similarly to the photoelectric effect described in the previous section, the photoconductivity has a threshold photon energy, and a corresponding threshold wavelength, λ_{th}. For an intrinsic semiconductor, λ_{th} is determined by the condition that the photon energy should be larger than the bandgap energy. According to the wavelength to be detected, different semiconductors, such as germanium, silicon, lead sulfide, indium antimonide can be used. In the case of photoconductive detectors operating at mid-infrared, low temperature operation is necessary, in order to avoid that the thermal excitation of electron-hole pairs obscures the photoexcitation. Instead of using small-bandgap semiconductors, the detection of mid-infrared radiation is more frequently performed using doped photoconductors. As an example, p-type germanium doped with Zn, cooled at 4 K, can detect mid-infrared radiation up to wavelengths of about 40 μm.

Photodiodes. A very important category of photodetectors is represented by photodiodes. A photodiode is a type of detector capable of converting light into either current or voltage, depending upon the mode of operation. As discussed in Sect. 5.1, a large internal electric field exists in the depletion layer of a p-n junction. If an electron-hole pair is generated by photon absorption within this layer, then the two charge carriers rapidly drift in opposite directions, the electron toward the n region

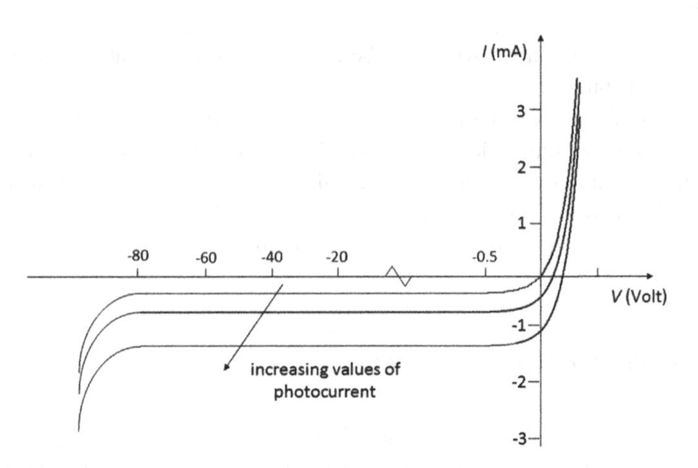

Fig. 5.15 Current-voltage characteristic of a photodiode. Under illumination the characteristic shifts downward by an amount proportional to the light intensity

and the hole toward the p region. If the device is left in an open circuit configuration, then an external potential appears between the p and n regions. This is known as the photovoltaic effect. The usual way to operate the detector is to apply a reverse voltage across the junction. The current-voltage characteristic expressed by (5.3), as modified by photon absorption, becomes:

$$i = i_s\left[\exp\left(\frac{V}{V_o}\right) - 1\right] - i_{ph}, \tag{5.12}$$

where i_{ph}, the photogenerated inverse current, is given by (5.10). As shown in Fig. 5.15, the effect of illumination is that of vertically translating the current-voltage characteristic. The internal resistance of a photodiode is large, and thus it behaves essentially as a current generator.

Many photodiodes use a structure in which an intrinsic region is interposed between the n and p regions. These photodiodes, known as p-i-n photodiodes exhibit a larger sensitive volume, and faster response time, with respect to p-n photodiodes. Furthermore the p-i-n structure has a width of the depletion region that is practically independent of the applied voltage.

For the detection of visible and near-infrared radiation silicon photodiodes are the most common choice, and can have a large efficiency: $\eta = 0.4$–0.7. For the wavelength interval 1.0–1.6 μm, *InGaAs* photodiodes are mostly used.

Avalanche photodiodes. The current-voltage characteristics shown in Fig. 5.15 illustrate a feature that is not described by (5.12). When the reverse bias voltage is sufficiently large, a sudden increase in reverse current is observed. What happens in this region is that carriers crossing the depletion layer acquire sufficient energy to ionize lattice atoms by collision. The carriers produced in this way may, in turn, generate further ionizing collisions in a cascade. This mechanism, known as "avalanche breakdown", can be destructive if the reverse current is allowed to become too large.

An avalanche photodiode (APD) is a photodiode designed to operate under very high reverse bias, in order to exploit the internal amplification caused by the avalanche breakdown. The applied voltage is kept slightly below the breakdown voltage, so that the APD still behaves as a linear device, giving an output signal proportional to the incident optical power. The actual value of the breakdown voltage depends on the semiconductor material, device structure and temperature. It is typically around 50 V for InGaAs/InP APDs.

Some particular types of APDs can reach such a high gain that they become single-photon detectors, a kind of solid-state photomultiplier tube, but requiring a supply voltage more than one order of magnitude lower than a real photomultiplier tube. A possible drawback is that the gain G is rather sensitive to temperature and aging.

5.5.3 CCD Image Sensors

It is possible to convert an optical image into a digital sequence of electrical signals using an array of photodetectors. The invention of the charge-coupled-device (CCD) structure represented a real revolution for image recording and transmitting. Boyle and Smith were awarded the Nobel prize in Physics in the year 2009 for this invention.

The photosensitive surface is divided into a two-dimensional array of detectors. The single detector, known as "pixel", a term coming from the contraction of "picture element", is a metal-oxide-semiconductor (MOS) capacitor. In visible CCD cameras the semiconductor is doped-silicon. A positive voltage applied between metal and silicon traps the photogenerated electrons at the silicon surface. The number of trapped electrons is proportional to the integral of the optical power incident on the pixel during the measurement interval. The electric charge relative to each pixel is amplified, sequentially read-out, and then recorded in digital form. Two-dimensional arrays typically consist of 1024×1024 pixels. The pixel size is of the order of 10 μm. Colored images can also be digitally converted by simultaneously recording images in the three fundamental colors. Infrared CCD cameras, based on *III–V* or *II–VI* semiconductors are also commercially available.

5.6 Electro-Absorption Modulators

An alternative to the electro-optic modulators described in Chap. 4 are the electro-absorption modulators based on semiconductors. The idea is to exploit the change in bandgap energy induced by an external electric field, an effect known as the Franz-Keldish effect. To ensure that the semiconductor is transparent to the optical signal, the wavelength λ of the signal to be modulated must be slightly larger than

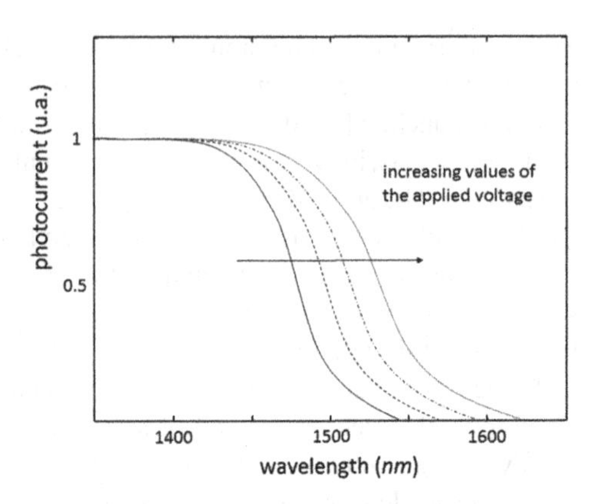

Fig. 5.16 Qualitative description of electro-absorption; modification of the wavelength dependence of the absorption coefficient (proportional to the photocurrent) at increasing applied fields

λ_g, so that the photon energy is less than E_g. The applied electric field induces a decrease of E_g, which is equivalent to an increase of λ_g, as shown in Fig. 5.16. If the new value of E_g is smaller or equal to the photon energy, then the modulated semiconductor absorbs the light beam. Therefore a time-dependent electric field induces an amplitude modulation of the optical field. Analogously, electro-absorption switching can be realized.

Electro-absorption is different from the electro-optic effects described in Chap. 4, because the effect of the applied field is to modify the imaginary part of the refractive index, instead of the real part. The change of E_g is rather small: typically, a field of about 10^7 V/m is needed to induce a shift of bandgap energy of 20 meV.

An interesting aspect is that it is possible to design a monolithic structure in which the electro-absorption modulator is integrated with a laser, to act as an external modulator, or inside the laser cavity to act as an intra-cavity modulator for active mode-locking.

The electro-absorption modulators used in optical communications (at wavelengths around 1.55 μm) are based on the quaternary compounds *InGaAsP* or *InGaAlAs* grown on an *InP* substrate. These modulators often have the form of a waveguide, with the electric field applied in a direction perpendicular to the propagation direction of the light beam. Multiple quantum well structures can be used to obtain high modulation speed and large extinction ratios, with low drive voltages. As an example, with a waveguided structure of length 1.5 mm and drive voltage of 11 V, an extinction ratio of 30 (about 15 dB) and a maximum modulation frequency of 20 GHz can be obtained.

Problems

5.1 Consider a semiconductor laser made of a Fabry-Perot cavity in which the optical feedback is only due to the refractive-index mismatch at the air-semiconductor interface. Assuming that the cavity length is 1 mm and that the refractive index of the active medium is 3.5, calculate: (i) the reflectance of the interfaces; (ii) the small-signal gain coefficient g (cm^{-1}) at threshold; (iii) the frequency separation between two adjacent longitudinal modes of the cavity.

5.2 Consider a $Ga_xIn_{1-x}N$ semiconductor laser. Assuming that the bandgap energy E_g is a linear function of the gallium fraction x and using the data of Table 5.1, calculate the value of x to be chosen for obtaining an emission wavelength of 420 nm.

5.3 A current is injected into a *GaInAsP* semiconductor laser of bandgap energy $E_g = 0.92$ eV and refractive index $n = 3.5$ such that the difference in Fermi levels $E_{fn} - E_{fv}$ (see Eq. (5.4)) is 0.95 eV. If the laser cavity has length $L = 300$ μm, determine the maximum number of longitudinal modes that can oscillate.

5.4 Calculate the bandwidth $\Delta\lambda$ of the light emitted by a blue $Ga_{0.45}In_{0.55}N$ LED at room temperature, by assuming that the bandgap energy E_g of the semiconductor is a linear function of the gallium fraction x and using the data of Table 5.1.

5.5 A photodiode with an efficiency of $\eta = 0.1$ is illuminated with 10 mW of radiation at 750 nm. Calculate the voltage output across a load resistance of 50 Ω.

5.6 A ten-dynode photomultiplier tube with an efficiency of $\eta = 0.1$ is illuminated with 1 μW of radiation at 750 nm. Assuming that each incident electron generates 4 secondary electrons at all dynodes, calculate the voltage output across a load resistance of 50 Ω.

Chapter 6
Optical Fibers

Abstract This chapter describes the waveguiding properties of optical fibers, with emphasis on the single-mode fibers used in optical communications. After examining modal behavior and attenuation processes, the dispersive propagation of ultrashort pulses is treated in some detail. A section is dedicated to the description of several different types of fibers and their properties, followed by a presentation of some fiber components, such as fiber couplers and fiber mirrors, which play an important role in the setup of fiber sensors and active fiber devices. The last two sections specifically deal with the active devices, such as fiber amplifiers and fiber lasers.

6.1 Properties of Optical Fibers

As discussed in Sect. 2.6, optical waveguides can confine an optical beam indefinitely, avoiding diffraction effects. An optical fiber is a cylindrical waveguide of one transparent material, clad by a second transparent material. It has important applications in many areas, primarily optical communications. The basic structure in shown in Fig. 6.1. Both regions are typically made of glass, with the core glass having a higher refractive index than the cladding. Within the frame of geometrical optics the confinement effect can be explained by considering a ray which passes through the center of the guide and is incident on the core-cladding interface at an angle larger than the limit angle. The ray is then able to travel along the guide by means of a sequence of total internal reflections.

The glasses used in optical fibers are made predominantly by silica, SiO_2, with the addition of small amounts of other oxides, such as germania (GeO_2) and alumina (Al_2O_3). Doping silica with germania or alumina raises the refractive index, and so a silica-based optical fiber could be made from a germania-doped silica core with a pure silica cladding. Silica is an ideal material because it can be purified to a high degree to become extremely transparent.

The fabrication of optical fibers is a two-stage process. In the first stage, a vapor-deposition method is used to make a cylindrical preform with the desired refractive

V. Degiorgio and I. Cristiani, *Photonics*, Undergraduate Lecture Notes in Physics,
DOI 10.1007/978-3-319-20627-1_6

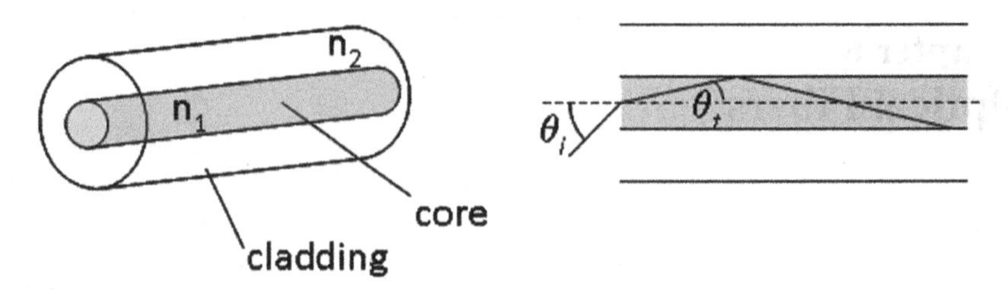

Fig. 6.1 The basic structure of an optical fiber

index profile and the relative core-cladding dimensions. The preform is a glass rod with a diameter between 1 and 10 cm and roughly 1 m length. Glass fibers are fabricated by pulling from the preform in a fiber-drawing tower. The fiber drawing process, as illustrated in Fig. 6.2, begins by feeding the glass preform into the drawing furnace. The drawing furnace softens the end of the preform to the melting point. When the preform is heated close to the melting point, a thin fiber can be pulled out of the bottom of the preform. During the pulling process, the fiber diameter is held constant by automatically adjusting the pulling speed with a feedback system. Fibers are pulled at a rate of 10–20 m/s. The pulling process preserves the transverse refractive index profile of the preform, by simply shrinking the diameter from a few centimeters to typically 125 μm. The fiber from a single preform can be many kilometers long. Before the fiber is wound up, it usually receives a polymer coating for mechanical and chemical protection.

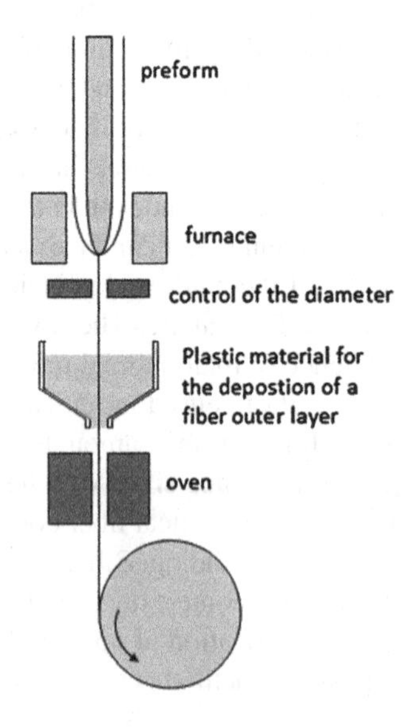

Fig. 6.2 Fiber drawing process

6.1.1 Numerical Aperture

A ray that is incident on the fiber front-face at an angle θ_i is refracted at an angle θ_t given by the Snell's law of (2.25):

$$n_o \sin\theta_i = n_1 \sin\theta_t \tag{6.1}$$

where n_o is the refractive index of the medium in which the fiber is immersed, and n_1 is the refractive index of the core glass. The transmitted beam hits the core-cladding interface at an incidence angle $\pi/2 - \theta_t$, and is totally reflected if this angle is larger than the limit angle of $\arcsin(n_2/n_1)$, where n_2 is the refractive index of the cladding glass. The incident ray is trapped inside the fiber if $\theta_i \leq \theta_{\max}$, where the maximum acceptance angle $\theta_{\max}$ is determined by the condition:

$$\pi/2 - \theta_t = \arcsin(n_2/n_1) \tag{6.2}$$

By using (6.1) and (6.2) it is found that $\theta_{\max}$ is given by:

$$\sin\theta_{\max} = \frac{\sqrt{{n_1}^2 - {n_2}^2}}{n_o} \tag{6.3}$$

The quantity $n_o \sin\theta_{\max}$ is known as the "numerical aperture" of the fiber, usually designated as NA, the same symbol used for the numerical aperture of the lens. Although the two definitions are somewhat different, the physical meaning is the same: the numerical aperture is a measure of the light-gathering capacity of the optical component. As an example, taking $n_1 = 1.48$, $n_2 = 1.46$, $n_o = 1$, one finds $NA = 0.24$. In this case the maximum acceptance angle is $\theta_{\max} = 14°$.

From now on it will be always assumed that $n_o = 1$. Sometimes it is also useful to introduce the parameter Δ, defined as:

$$\Delta = \frac{{n_1}^2 - {n_2}^2}{2{n_1}^2} \approx \frac{n_1 - n_2}{n_1} \tag{6.4}$$

6.1.2 Modal Properties

The optical fiber described in the preceding section is known as a "step-index fiber", because the refractive index profile presents an abrupt change at the core-cladding interface. It is also possible to guide a light beam using a refractive index profile that decreases smoothly inside the core from the center to the cladding. These fibers are known as "graded-index" fibers. Their refractive index profile can be generally expressed in the form:

$$n(r) = n_2 + (n_1 - n_2)\left[1 - (r/a)^\gamma\right], \tag{6.5}$$

where a is the core radius and r is the radial distance. The validity of (6.5) is limited to the interval $0 \leq r \leq a$. If the exponent γ tends to infinity, (6.5) reproduces a step-index profile. When $\gamma = 2$, there is a parabolic profile.

The propagation of optical waves along fibers must be treated, following the general scheme outlined in Sect. 2.6, by using the Helmholtz equation and the boundary conditions at the core-cladding interface. Since the geometry is two-dimensional, the modes are characterized by two integer numbers, m and l. Using the cylindrical coordinates z, r, and θ, the electric field of a generic mode, written in scalar form, is given by:

$$E_{lm}(r, \theta, z, t) = AB_{lm}(r, \theta)\exp[i(\beta_{lm}z - \omega t)], \tag{6.6}$$

where β_{lm} is the propagation constant and $B_{lm}(r, \theta)$ is the transverse mode profile, which is independent of the propagation distance z.

The main results of the theoretical treatment are summarized as follows. The propagation properties are described by a set of so-called "linearly polarized" (LP_{lm}) modes, where an LP_{lm} mode has m field amplitude maxima along the radial coordinate and $2l$ maxima along the circumference. The lowest order mode, i.e. the LP_{01} mode, has a bell-shaped field amplitude distribution that is well approximated by a Gaussian function:

$$B_{01}(r, \theta) = B_o\exp\left(-\frac{r^2}{w^2}\right), \tag{6.7}$$

where w is known as the mode field radius. It has been seen in Sect. 1.4.2 that freely propagating Gaussian waves have z-dependent parameters. In contrast, the Gaussian wave propagating inside the fiber has constant radius and a fixed planar wavefront.

While there is always a solution for the fundamental mode LP_{01}, the existence of higher-order modes depends on the fiber parameters. It is useful to introduce a dimensionless parameter, V, known as the normalized frequency, defined as:

$$V = \frac{2\pi a}{\lambda}\sqrt{n_1{}^2 - n_2{}^2} = \frac{2\pi a}{\lambda}NA. \tag{6.8}$$

It can be shown that, if V is smaller than a critical value, V_c, then the fiber can support only the fundamental mode. In the case of a "step-index" fiber, the critical value is: $V_c = 2.405$. In the more general case of a profile following the law (6.5), V_c is given by:

$$V_c = 2.405\sqrt{1 + \frac{2}{\gamma}}. \tag{6.9}$$

As an example, for $\gamma = 2$, $V_c = 3.512$. Above V_c the second mode, LP_{11}, can also propagate. As long as V increases, more and more higher-order modes can propagate. When $V \gg 1$, the number of modes, N, that can propagate is given by

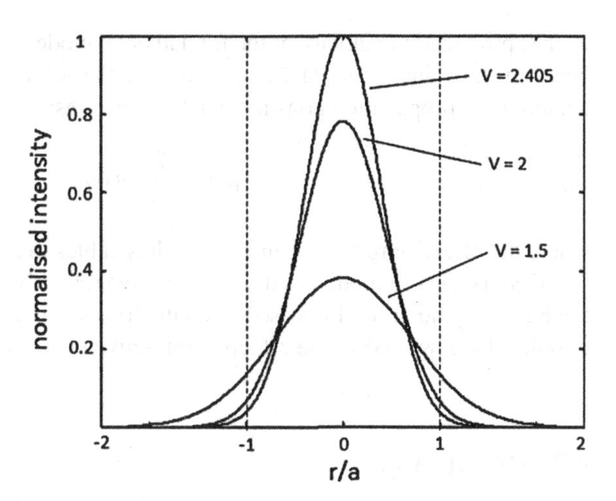

Fig. 6.3 Intensity of the fundamental mode plotted as a function of the ratio between the distance from the axis and the core radius, at different values of the normalized frequency V

$$N = \frac{V^2}{2} \tag{6.10}$$

At fixed fiber parameters, the condition $V = V_c$ defines a cut-off wavelength:

$$\lambda_c = \frac{2\pi a}{V_c}\sqrt{n_1{}^2 - n_2{}^2} = \frac{2\pi a}{V_c} NA. \tag{6.11}$$

At wavelengths $\lambda \geq \lambda_c$ the fiber behaves as a single-mode waveguide.

The intensity profile of the fundamental mode is shown in Fig. 6.3 for three values of V. The optical power is mainly traveling inside the core, but there is also an exponential tail (the evanescent wave described in Sect. 2.2.7) extending inside the cladding. If V is larger, then the confinement inside the core is more effective because the $n_1 - n_2$ difference is larger. The dependence of the beam radius w on V is described by the empirical law:

$$w = a(0.65 + 1.62\ V^{-1.5} + 2.88\ V^{-6}) \tag{6.12}$$

Single-mode step-index fibers are designed by keeping the value of V close to 2. Smaller values of V are avoided because the field amplitude extends increasingly into the cladding as V decreases. If the field amplitude is not negligible at the edge of the cladding, then the power losses may become significant.

The single-mode condition imposes small core radius values. As an example, a single-mode fiber, operating at $\lambda = 1550$ nm, and designed to have $V = 2$, can be made by taking $a/\lambda = 2.5$ and $NA = 0.12$. The core radius is only 3.9 μm. The cut-off wavelength of such a fiber is 1290 nm.

The propagation velocity of the fundamental mode is expected to be intermediate between the values c/n_1 and c/n_2, because the wave extends over both core and cladding. Its propagation constant can be written as:

$$\beta_{01} = \frac{2\pi}{\lambda} n_{eff}, \tag{6.13}$$

where the effective refractive index n_{eff} has values in the interval $n_1 \geq n_{eff} \geq n_2$.

Of course, every mode has its own effective refractive index. As a general rule, if the indices l and m are increased, then the transverse distribution is more extended into the cladding, and so the effective refractive index is expected to decrease.

6.2 Attenuation

The beam propagation inside the optical fiber is unavoidably lossy, so that the optical power decreases with the propagation distance z. Assuming linear losses, the optical power decays following an exponential law:

$$P(z) = P_o \exp(-\alpha z), \tag{6.14}$$

where P_o is is the power launched into the fiber and α is the attenuation coefficient, already defined in Sect. 2.1. The unit commonly used to characterize loss in fibers is decibel per kilometer (dB/km). Denoting by α_o the attenuation coefficient measured in dB/km, it is found:

$$\alpha_o = \frac{10}{z} \log_{10}\left[\frac{P_o}{P(z)}\right]. \tag{6.15}$$

If α is expressed in units of km^{-1}, then the numerical relation between the two coefficients is:

$$\alpha_o = 10\,\alpha\,\log_{10} e = 4.34\,\alpha \tag{6.16}$$

From (6.15) it is found that the propagation distance at which the incident power is halved, $z_{1/2}$, is given by:

$$z_{1/2} = \frac{10 \log_{10} 2}{\alpha_o} \approx \frac{3}{\alpha_o}. \tag{6.17}$$

As an example, if $\alpha_o = 0.15$ dB/km, then $z_{1/2} = 20$ km.

Fiber losses are due to two effects, light scattering and absorption. Scattering arises because there are submicroscopic random inhomogeneities in density and composition inside the glass, which give rise to local fluctuations in the refractive index. The inhomogeneities originate from thermal fluctuations in the melt of the silica that are then frozen in, as the glass cools. Light scattering from submicroscopic scatterers, known as Rayleigh scattering, gives a wavelength-dependent attenuation coefficient that scales as λ^{-4}.

Silica glasses have an absorption band in the ultraviolet, due to electronic transitions, and absorption bands in the mid-infrared, due to vibrational transitions. At wavelengths greater than about 1.6 μm, absorption losses become predominant with respect to scattering losses. The combination of scattering and absorption losses gives rise to the attenuation curve shown in Fig. 6.4, which has a minimum of ≈0.15 dB/km at about 1.55 μm. Indeed, long-distance optical communication systems use semiconductor lasers emitting a 1.55 μm beam, as the carrier that is modulated and transmitted through the optical fiber.

The attenuation curve of Fig. 6.4 also shows a secondary minimum around 1.3 μm, due to the presence of an absorption peak at about 1.4 μm. This peak, due to vibrational transitions of the O–H bond, occurs because of small quantities of water trapped inside the glass during the fiber fabrication process. This absorption peak no longer exists in recent fibers, because of improvements in the purification process. However, several networks using fibers with water impurities are still installed and in use.

Historically, the first optical-communication experiments, operating at short distances, were performed at 850 nm. In optical-communication jargon, the wavelength region around 850 nm was called the first attenuation window. The region around 1.3 μm is called the second attenuation window, and the one around 1.55 μm the third attenuation window.

All the glasses made by different combinations of oxides offer similar infrared absorption bands, so that the attenuation behavior shown in Fig. 6.4 is not much changed by varying the oxide composition. It is, however, possible to shift the vibrational absorption to longer wavelengths by using fluoride-based glasses instead of oxides. The so-called "ZBLAN" glasses, made from a mixture of zirconium, barium, lanthanum, aluminum, and sodium fluorides, offer attenuation minima around 3 μm.

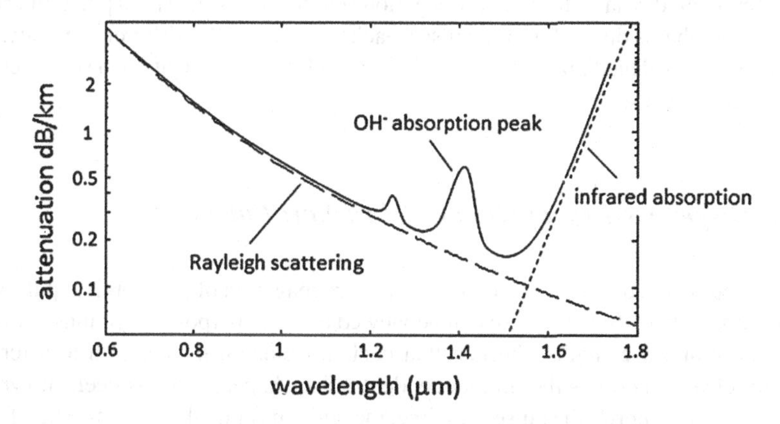

Fig. 6.4 Attenuation coefficient of a silica fiber versus wavelength

The minimum value of the attenuation coefficient is one order of magnitude lower than for silica glasses.

In addition to intrinsic losses, there are also fiber losses due to the departure from the ideal straight line configuration. In a bent fiber, the incidence angles at the core-cladding interface change, and so the condition of total internal reflection is no longer satisfied, meaning that the optical power is only partially trapped inside the fiber. In general, bending losses are larger for those modes that extend more into the cladding. Typically bending losses become important when bend radii are of the order of millimeters.

6.3 Dispersion

Optical dispersion, as defined in Sect. 2.1, is related to the dependence of the refractive index of a material on the wavelength of the propagating beam. As discussed in Sect. 6.1.2, a mode propagating in an optical fiber has an effective refractive index, which is a type of weighted average between n_1 and n_2. Since both materials are dispersive, one might think that the dispersion of n_{eff} is determined by that of the core and cladding materials, to a first approximation. The situation, however, is more complex, because the parameter V is changed by changing the wavelength. Therefore, the transverse field distribution is modified. As a consequence the relative weighting of n_1 and n_2 varies with λ. The conclusion is that the single-mode dispersion is a combination of both a material effect and a waveguide effect. The waveguide dispersion is very sensitive to the shape of the radial profile of the refractive index, such that it may significantly change on going from a step-index to a graded-index fiber.

An intermodal dispersion occurs in multimode fibers, due to the differences in effective refractive indices of the various modes. Even if the wavelength is the same, the different modes have different propagation velocities. A single light pulse entering an N-mode fiber is divided into N pulses, each traveling with a different velocity. At the output a broadened pulse is observed, or, for a longer propagation, a sequence of N separated pulses.

6.3.1 Dispersive Propagation of Ultrashort Pulses

In this section, the discussion is focused on the propagation of short optical pulses in single-mode fibers. Since a pulse can be viewed as the superposition of many waves with different wavelengths, the fact that each wavelength travels with a different velocity clearly modifies the time-dependent pulse-shape. As it has been shown in Sect. 3.11.3, the shorter the pulse is in time, the wider its optical spectrum. Therefore, one should expect that dispersion effects will be largest for the shortest pulses.

In order to describe pulse propagation it is useful to express the propagation constant $\beta(\omega) = (\omega/c)n_{eff}(\omega)$ as a power series around the central pulse frequency ω_o:

$$\beta(\omega) = \sum_{k=0}^{\infty} \frac{d^k\beta}{d\omega^k} \frac{(\omega - \omega_o)^k}{k!}, \tag{6.18}$$

where the derivatives $d^k\beta/d\omega^k$ are evaluated at ω_o.

Assuming that the pulse spectral width, $\Delta\omega$, is much smaller than ω_o, the power expansion can be truncated at the quadratic term:

$$\beta(\omega) = \beta_o + \beta_1(\omega - \omega_o) + \frac{1}{2}\beta_2(\omega - \omega_o)^2, \tag{6.19}$$

where $\beta_o = \beta(\omega_o)$,

$$\beta_1 = \frac{d\beta}{d\omega}, \tag{6.20}$$

and

$$\beta_2 = \frac{d^2\beta}{d\omega^2}. \tag{6.21}$$

The electric field of the input pulse, entering the single-mode fiber at $z = 0$, is given by

$$E(t, 0) = A(t, 0)\exp(-i\omega_o t). \tag{6.22}$$

For simplicity, the term $B_{01}(r)$, which describes the transverse distribution of the field amplitude and is invariant during the propagation, has been omitted.

The mathematical treatment of dispersive propagation becomes simpler if just one specific pulse shape is considered, such as the Gaussian shape described by (3.77). This has a Fourier transform given by (3.80).

In the frequency domain, the effect of propagation may be regarded as a frequency-dependent phase delay, given by $\phi(\omega, z) = \beta(\omega)z$. In this way, the Fourier transform of the field at z is simply given by:

$$E(\omega, z) = E(\omega, 0)\exp[i\beta(\omega)z] \tag{6.23}$$

By substituting (6.19) and (3.80) inside (6.23) one obtains:

$$E(\omega, z) = \sqrt{2\pi}\,\tau_i A_o \exp(i\beta_o z)\exp\left[-({\tau_i}^2 - i\beta_2 z)\frac{(\omega - \omega_o)^2}{2} + i\beta_1 z(\omega - \omega_o)\right]. \tag{6.24}$$

By using (2.9), it is possible to return to the time domain, in order to recover the expression for the time-dependent electric field at the generic coordinate z:

$$E(t,z) = \frac{1}{2\pi}\int_{-\infty}^{\infty} E(\omega,z)\exp(-i\omega t)d\omega$$
$$= \frac{A_o}{\sqrt{1 - i\beta_2 z/\tau_i^2}}\exp[-i(\omega_o t - \beta_o z)]\exp\left[-\frac{(t-\beta_1 z)^2(1 + i\beta_2 z/\tau_i{}^2)}{2\tau_i{}^2(1+\beta_2{}^2 z^2/\tau_i{}^4)}\right] \tag{6.25}$$

It is useful to separately highlight the behavior of the field amplitude and phase by writing the electric field as:

$$E(t,z) = A(t,z)\exp\{-i[\omega_o t - \beta_o z + \phi(t,z)]\}, \tag{6.26}$$

where:

$$A(t,z) = \frac{A_o}{(1+\beta_2{}^2 z^2/\tau_i{}^4)^{1/4}}\exp\left[-\frac{(t-\beta_1 z)^2}{2\tau_i{}^2(1+\beta_2{}^2 z^2/\tau_i{}^4)}\right] \tag{6.27}$$

and

$$\phi(t,z) = \frac{(t-\beta_1 z)^2 z\beta_2/\tau_i{}^2}{2\tau_i{}^2(1+\beta_2{}^2 z^2/\tau_i{}^4)} - \frac{\arctan(\beta_2 z/\tau_i{}^2)}{2}. \tag{6.28}$$

Equation (6.27) shows that the time-dependent pulse amplitude remains Gaussian during the dispersive propagation. The pulse intensity, $I(z,t)$, proportional to the modulus square of (6.27), reads:

$$I(t,z) = \frac{I_o}{\sqrt{1+(z/L_D)^2}}\exp\left[-\frac{(t-\beta_1 z)^2}{\tau_i{}^2(1+z^2/L_D{}^2)}\right], \tag{6.29}$$

where I_o is the peak intensity of the input pulse. Equation (6.29) contains the dispersion length L_D, defined as:

$$L_D = \frac{\tau_i{}^2}{|\beta_2|} \tag{6.30}$$

According to (6.29), the intensity observed at the coordinate z is maximum at time $t_p = \beta_1 z$. This means that the pulse peak is propagating with a velocity, known as the "group velocity", which is given by:

$$u_g = \beta_1{}^{-1}. \tag{6.31}$$

The quantity β_1 can be written as:

$$\beta_1 = \frac{1}{c}\left(n_{eff} + \omega\frac{dn_{eff}}{d\omega}\right) = \frac{n_g}{c}, \tag{6.32}$$

where n_g is called group refractive index. As is the case for most transparent materials, the refractive index of silica glass, n_{gl}, is an increasing function of ω. Although n_{eff} does not coincide with n_{gl}, its frequency-dependence is similar, so that the derivative $dn_{eff}/d\omega$ is positive. Therefore it can be quite generally said that the group velocity is smaller than the phase velocity c/n_{eff}, according to (6.32).

Recalling that the pulse duration τ_p is related to τ_i by (3.81), the law describing the change of pulse duration upon propagation can be derived from (6.29):

$$\tau_p(z) = \tau_p(0)\sqrt{1 + \frac{z^2}{{L_D}^2}} \tag{6.33}$$

Equation (6.33) shows that L_D represents the propagation distance above which there is a significant increase of the pulse duration. While the pulse spectrum remains unchanged during propagation, the pulse duration grows with z. As a consequence the bandwidth-duration product is no longer given by (3.81), but becomes:

$$\Delta\nu_p\tau_p(z) = 0.441\sqrt{1 + \frac{z^2}{{L_D}^2}} \tag{6.34}$$

Since a lossless propagation has been considered, the pulse energy must be conserved. Indeed (6.29) shows that the intensity decreases in such a way as to keep the product between peak intensity and pulse duration constant during propagation.

The quantity β_2 is known as the "group velocity dispersion" (GVD) parameter, and is usually measured in units of ps^2/km. At the optical-communication wavelength, $\lambda = 1.55\ \mu$m, it is found that $\beta_2 = -20$ ps^2/km for a typical single-mode fiber. It is instructive to numerically investigate the effect of pulse duration. For a 20-ps pulse, (6.30) gives $L_D = 20$ km. After a propagation length of 20 km, the pulse duration is increased by a factor of $\sqrt{2}$. For a 1-ps pulse, L_D is much shorter, only 50 m. Here, the pulse duration is increased by a factor of 400, after propagating 20 km. This clearly shows that the stretching effect due to dispersion plays a very important role for ultrashort pulses.

In the area of optical communications, it is customary to use the dispersion parameter D, instead of β_2. This is defined as:

$$D = \frac{d\beta_1}{d\lambda} = -\frac{2\pi c\beta_2}{\lambda^2} \tag{6.35}$$

D is measured in units of ps/(nm km). It represents the group delay per wavelength (in nm) and propagation length (in km). At $\lambda = 1.55\ \mu$m, it is found that $D =$ 17 ps/(nm km) for a typical single-mode fiber. The so-called "normal" group-velocity

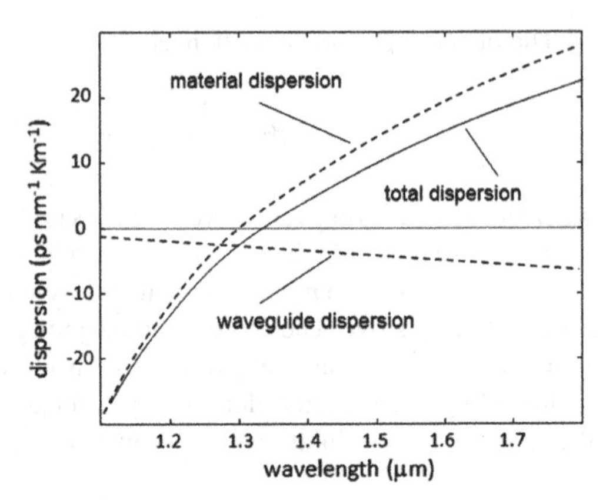

Fig. 6.5 Material and waveguide dispersion in a single-mode silica fiber with a parabolic index profile

dispersion regime corresponds to the situation in which D is negative (and β_2 is positive). "Anomalous" dispersion corresponds to a positive D (and negative β_2).

Figure 6.5 shows the behavior of D as a function of λ for a single-mode silica fiber with a parabolic index profile. The behavior is mainly determined by the material dispersion, with only a slight modification introduced by waveguide dispersion. The zero-dispersion wavelength is about 1.3 μm. The fiber dispersion is normal at wavelengths less than 1.3 μm, and anomalous for longer wavelengths.

6.4 Fiber Types

In this section a review of the main types of fibers available is given. A first distinction must be made between single-mode fibers and multimode fibers. As discussed in Sect. 6.1.2, single-mode fibers have $V \approx 2$ and a core diameter of ≈8 μm, for transmission at 1.55 μm. Multimode fibers have a large V and core diameter typically around 50 μm. The numerical value of the cladding diameter is not critical as long as it is large enough to confine the fiber modes entirely. The cladding diameter of single-mode fibers is usually standardized to 125 μm. Multimode fibers have a larger NA, which makes it much easier to couple optical signals into the fibers. They are also cheaper and can carry a higher optical power. Intermodal dispersion limits their utility for pulse transmission to very short distances.

Low-dispersion fibers. Optical communications require low-dispersion fibers. There are available fibers in which the zero-dispersion wavelength is shifted from 1.3 μm to about 1.55 μm, by using the w-shaped profile shown in Fig. 6.6. In this type of fiber, known as "dispersion shifted" (DS) fiber, the waveguide dispersion almost exactly compensates the material dispersion at 1.55 μm, as shown in Fig. 6.7. Other fibers, known as "dispersion-flattened" fibers, are designed to have a small non-zero dispersion over a wide wavelength interval around 1.55 μm, as shown in Fig. 6.7. In the area of optical communications, as it will be seen in Chap. 8, it is also important

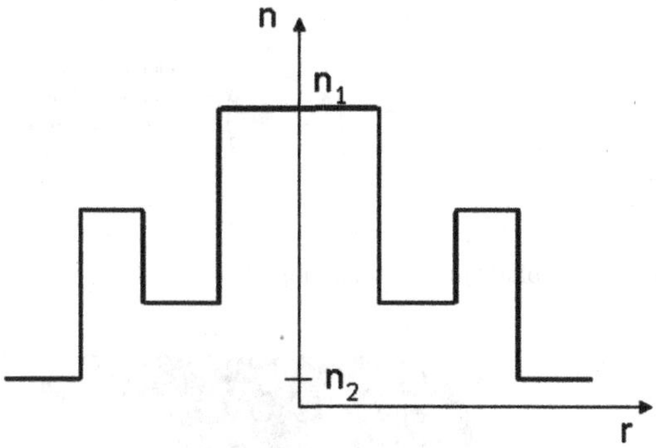

Fig. 6.6 Refractive index profile in a "dispersion-shifted" fiber

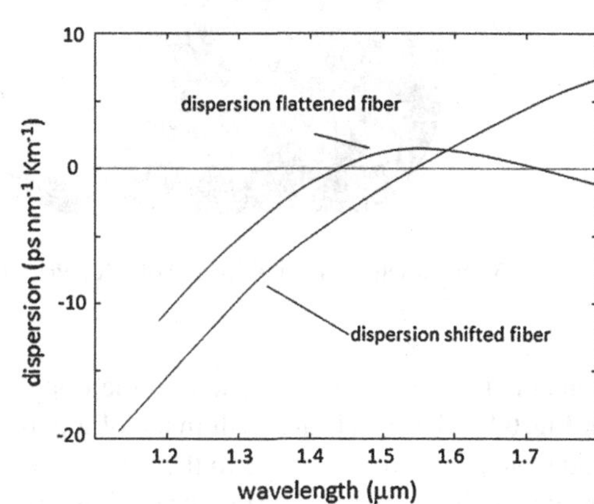

Fig. 6.7 Group velocity dispersion versus λ for a "dispersion-shifted" and a "dispersion flattened" fiber

to have fibers with large normal dispersion ($D \approx -60$ ps nm^{-1} km^{-1}), available for use in dispersion compensation. A short stretch of these fibers (known as "dispersion compensating" DIS-CO) can compensate for the dispersion accumulated over a long propagation span in an SMR fiber.

Polarization-maintaining fibers. An aspect of waveguiding that has not yet been touched upon in this chapter is that of signal polarization. The fundamental mode LP_{01} is a linearly polarized mode, but its polarization direction is completely arbitrary as long as the fiber has perfect cylindrical symmetry about its axis. In fact, real fibers invariably have some slight local asymmetry, either intrinsic or due to their environment, which creates a small birefringence varying randomly along their length. As a result, the initial polarization is lost after some tens of meters, so that it becomes impossible to predict the polarization state of the output signal. In the case of pulse propagation, the random changes in the polarization can induce a pulse broadening. This dispersion effect, known as "polarization-mode dispersion" (PMD), is usually smaller than the GVD effect. For some applications it is desirable that fibers transmit light without changing the polarization state. Such fibers are called "polarization-maintaining" fibers. One possibility is to break the cylindrical symmetry of the fiber by using an elliptical core. An alternative scheme is that of fabricating a preform that

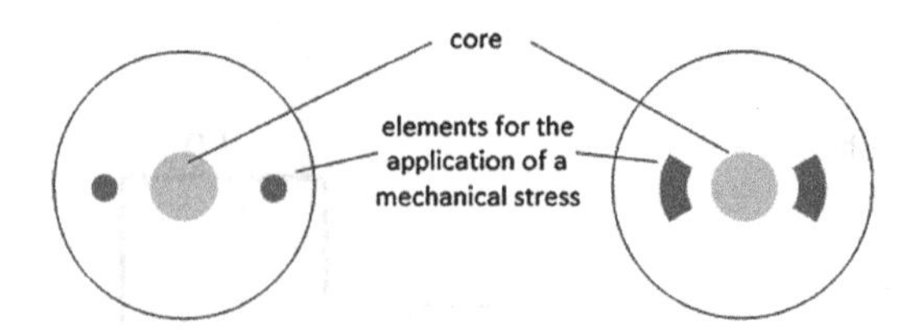

Fig. 6.8 Polarization-maintaining fibers

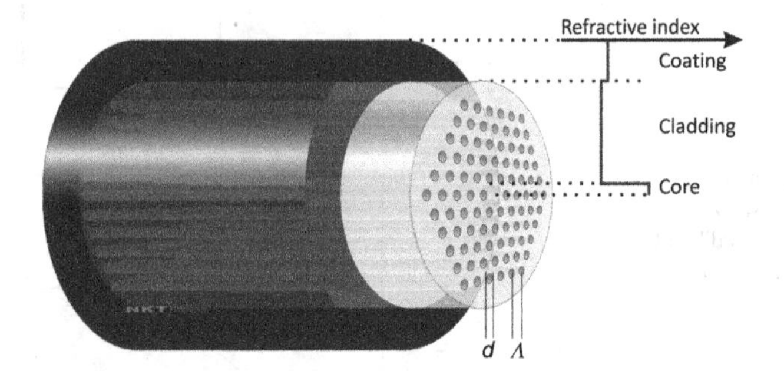

Fig. 6.9 Microstructured fiber. Courtesy of NKT Photonics

contains two glass rods, inserted on the opposite sides of the fiber core, as shown in Fig. 6.8. These rods are both made of a glass that is different from the cladding, and so mechanical stress due to their different thermal expansion coefficients arises inside the fiber structure during the cooling process. Because of stress-induced birefringence, the fiber becomes strongly birefringent, so that a linear polarization state parallel to one of the principal axes is transmitted without a change in polarization. This type of fiber is sometime called "panda" fiber.

Microstructured fibers. A new family of fibers, called microstructured fibers, contain cylindrical air holes running parallel to the fiber axis, as shown in Fig. 6.9. If the holes are organized in a regular periodic pattern, then the fiber is also called photonic-crystal fiber. These fibers are usually made of pure silica glass. Both the hole diameter and the minimum distance between holes are of the order of the wavelength of light. Microstructured fibers have higher losses and are more expensive than standard fibers. However, they are interesting for various applications, because their more complicated structure gives new degrees of freedom, offering performances not accessible using standard fibers. Once λ is fixed, there is a very limited flexibility regarding the choice of core diameter of standard single-mode fibers. In contrast, microstructured fibers can be designed for single mode operation with a large mode-area (mode diameter larger than 10 μm) or a very small mode-area (mode diameter around 1 μm). Large mode-areas are interesting for transmitting high optical powers, whereas small mode-areas produce intensities high enough to enhance nonlinear effects, such as the optical Kerr effect and other effects that will be described in the following chapter. The dispersion properties of microstructured fibers can also be drastically different from those of standard fibers. In particular, the zero-dispersion wavelength can be positioned anywhere in the visible or near infrared.

Plastic optical fibers. Fibers can also be made from polymers, by using, for instance a core of polymethyl methacrylate (PMMA, $n = 1.495$) and a cladding of fluoroalkyl methacrylate ($n = 1.402$). These fibers have a transparency window in the visible, between 500 and 700 nm, have large attenuation coefficients (50–100 dB/km) and also a large NA. They can be useful for sensors, for short-distance multimode data-transmission, and for architectural lighting.

6.5 Fiber-Optic Components

Several optical components can be made using fibers. They are useful for the manipulation of optical signals that are already traveling inside fibers, so that all-fiber setups can be realized. In this section attention is focused on two components, fiber couplers and fiber mirrors.

Couplers. Fiber couplers can couple light from one or more input fibers to one or more output fibers. Several fibers can be thermally fused, such that their cores come into contact over a length of some centimeters, as shown in Fig. 6.10. When the evanescent field of one fiber reaches the core of the other fiber, a fraction, F, of the power traveling in the first fiber is coupled into the other. An optical signal coming from fiber 1 can be progressively transferred to fiber 2. The power fraction that is coupled into fiber 2 depends on the length of the coupler, L, according to the law:

$$F = \sin^2(AL), \tag{6.36}$$

where the parameter A is an increasing function of λ. By varying L, fiber beam-splitters that have an arbitrary power ratio between the two output fibers can be designed. Fiber couplers introduce very low losses, typically 0.2–0.5 dB. Since the power transfer depends not only on the coupling length, but also on wavelength, it is possible to design couplers, which, for instance, entirely transfer a λ_1 signal from fiber 1 to fiber 2, while leaving a λ_2 signal in fiber 1. Such dichroic couplers are used in fiber amplifiers to combine the signal input and the pump wave, as it will be seen in Sect. 6.6. Their insertion loss may be very small (below 1 dB) for both inputs.

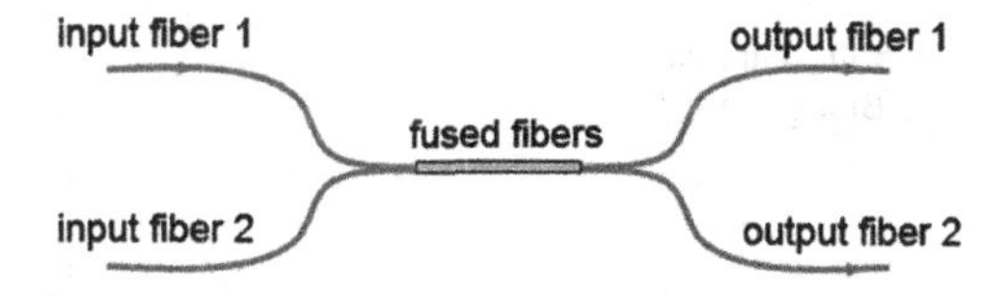

Fig. 6.10 Scheme of a fiber coupler

In optical-communication systems wavelength-sensitive couplers are used as multiplexers to combine several input signals with different wavelengths, or to separate different wavelengths.

Fiber Bragg gratings. An other type of fiber component is the wavelength-selective fiber mirror, which is conceptually similar to the multilayer mirrors discussed in Sect. 2.2.4. The idea is to create a periodic change of refractive index within the core. Such a phase grating, known as "fiber Bragg grating", has maximum reflectivity when the phase delay corresponding to the propagation along one period is equal to π. Therefore the wavelength of maximum reflectivity, λ_o, known as the Bragg wavelength, is given by the relation

$$\lambda_o = 2mn\Lambda, \tag{6.37}$$

where m is an integer, Λ is the periodicity of the grating and n is the refractive index of the core.

The maximum reflectivity is calculated using the same relations (2.55) and (2.56) given in Sect. 2.2.4 for multilayer mirrors. By putting $n_1 = n_2 = n_B = n$, and $n_A = n + \Delta n$ inside (2.56), $R_{\max}$ can be expressed as:

$$R_{\max} = \tanh^2\left(N\frac{\Delta n}{n}\right), \tag{6.38}$$

where N is the number of periods. The length, L, of the grating is typically of the order of a few millimeters, and so $N = L/\Lambda$ is of the order of 10^4. Even if Δn is small, a maximum reflectivity very close to 100 % can be obtained. As an example, taking $\Delta n = 2 \times 10^{-4}$, $N = 4 \times 10^4$, $n = 1.5$, it is found: $R_{\max} = 99.99\,\%$. The reflection bandwidth is typically narrower than 1 nm.

Fiber Bragg gratings are fabricated by exploiting the photosensitivity of germania-doped silica glasses to ultraviolet radiation. Radiation at a wavelength around 240 nm, usually coming from an excimer laser, induces a permanent change of refractive index in the core. The increase in refractive index due to UV illumination, Δn, is small, typically about 10^{-4}. However, this can be enhanced by forcing molecular hydrogen into the fiber core through a high-pressure low-temperature diffusion process. There are various methods to obtain a spatially periodic illumination of the core. The method shown in Fig. 6.11 uses a periodic phase mask, creating interference fringes with the required periodicity.

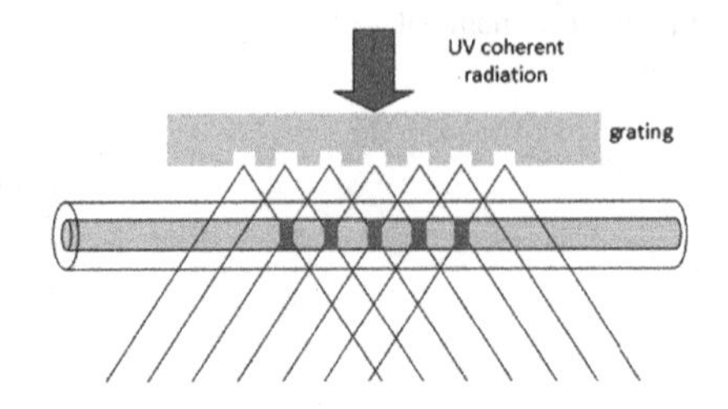

Fig. 6.11 Fabrication of a fiber Bragg grating

Fiber Bragg gratings are used as mirrors for fiber lasers, as will be seen in Sect. 6.7. They also have many applications in optical communication systems and in fiber sensors, as will be seen in Chap. 8. A recent interesting development is that of "long period gratings", in which the modulation period is larger than the wavelength. Inside these gratings there is a wavelength-selective transfer of optical power from core to cladding. The power displaced into the cladding is ultimately dissipated, and so the grating selectively suppresses certain wavelengths, acting as a filter.

Fiber gratings with a linear variation in the grating period (termed chirped gratings) are particularly useful for dispersion compensation. In these gratings, the Bragg wavelength varies with position, thereby broadening the reflected spectrum. A chirped grating has the property of adding dispersion, because different wavelengths are reflected from the grating subject to different delays. In this way, it is possible to achieve a very large group delay dispersion, sufficient for compensating the dispersion of a long span of transmission fiber, in a short length of fiber grating.

6.6 Fiber Amplifiers

An optical fiber can become the active medium for optical amplification, if the fiber core is doped with rare-earth ions, such as Nd^{3+}, Pr^{3+}, Er^{3+}, and Yb^{3+}. A strong motivation for the development of efficient fiber optical amplifiers came from the area of optical communications. In the case of long-distance links the attenuation due to fiber losses becomes large, and so it is necessary to periodically restore the power level of the traveling optical signals. Fiber amplifiers offer the best physical compatibility with fiber transmission systems.

Amplifiers based on erbium-doped fibers, which exhibit gain over a range of wavelengths centered on 1550 nm, were developed in the late 1980s. An important factor for the success of the erbium-doped fiber amplifier (EDFA) was the discovery that relatively high concentrations of erbium could be incorporated without ion clustering into the fiber core by co-doping with alumina.

The energy level scheme for the Er^{3+} ion in silica glass is presented in Fig. 6.12. By exploiting the long lifetime (about 12 ms) of the first excited level, optical gain can be obtained for the transition between this excited level and the ground state. The system then behaves like a three-level system, similar to that of the ruby laser described in Sect. 3.9. Because the ions are incorporated into a disordered solid matrix the energy levels are considerably broadened, and so the gain is spread over a wavelength bandwidth of about 40 nm. The peak gain is obtained at 1530 nm, as shown in Fig. 6.13.

Like the solid-state lasers described in Sect. 3.9, fiber amplifiers are optically pumped. Because of the constraints of the fiber geometry, they are longitudinally pumped, so that both pump and signal utilize the waveguiding effect of the fiber. This ensures an excellent overlap of the two waves over the entire length of the device, and allows the pump power to be completely absorbed and utilized.

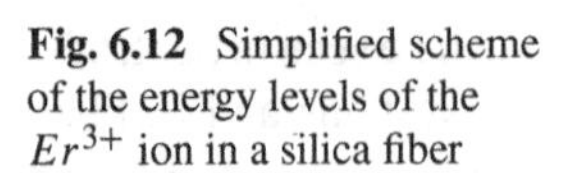

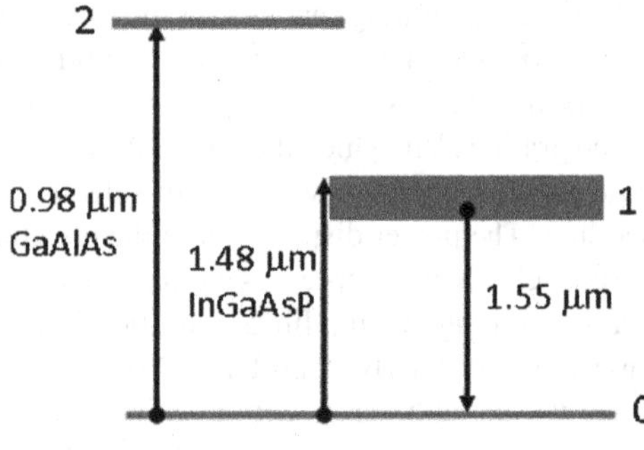

Fig. 6.12 Simplified scheme of the energy levels of the Er^{3+} ion in a silica fiber

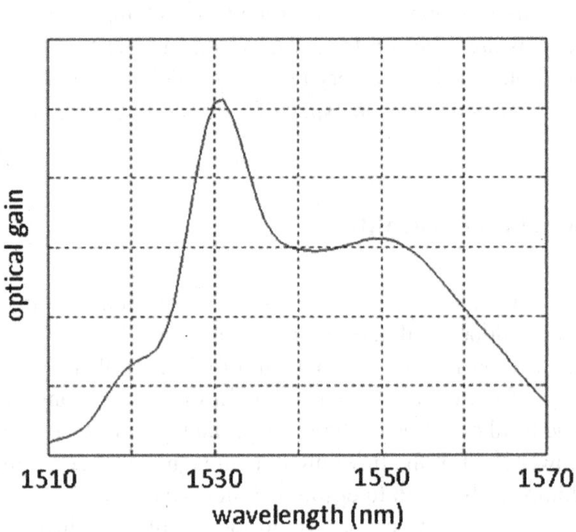

Fig. 6.13 Optical gain versus wavelength for a typical erbium-doped fiber amplifier

Optical pumping is carried out with semiconductor lasers, emitting at 980 nm or 1480 nm. In the former case, the ion excited at level 2 rapidly decays non-radiatively to the lower level of the band 1, so that a population inversion can be established between this lower level and level 0. In the latter case, the ion is excited to a level that is already inside band 1. By a rapid intraband non-radiative decay, the excitation is subsequently transferred to the lower level of the band 1.

A typical scheme of an erbium-doped fiber amplifier is shown in Fig. 6.14. The pump and signal beams are coupled into the fiber by using a dichroic fiber coupler of the type described in the previous section. An isolator is inserted after the doped fiber, in order to avoid back-reflections, which could have the undesired effect of transforming the amplifier into an oscillator. The residual pump power, possibly present at the output of the amplifier, is eliminated through a wavelength filter or a dichroic coupler. The pump light can also be injected in the opposite direction from the signal, or in both directions. The latter configuration provides a more uniform distribution of pump power along the length of the amplifier.

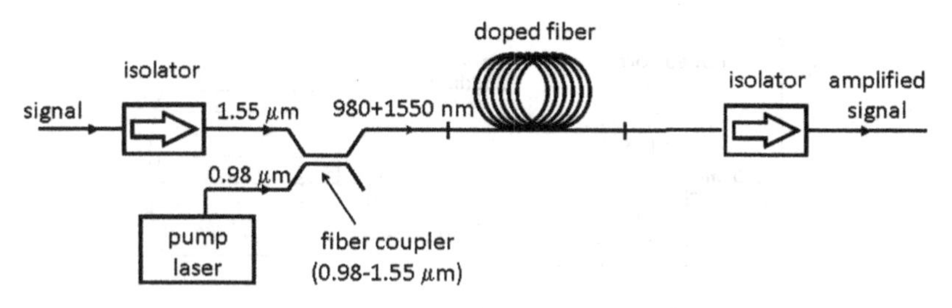

Fig. 6.14 Scheme of an EDFA

Once the type of doped fiber and the pump power are chosen, the parameter to be fixed is that of the length of the doped fiber, L. If the gain medium is too long, part of it will not be pumped because the pump power will be totally absorbed before it reaches the far end of the fiber. In a three-level system, such as an EDFA, the unexcited ions that remain in the ground state absorb signal photons, so that the un-pumped regions of the fiber have a detrimental effect on the amplifier performance. Considering that too short an amplifier also does not use the pump power very efficiently, it is clear that an optimum length for the fiber amplifier exists.

Typical parameters for an optical-communications amplifier are the following. The concentration of erbium atoms is $N_t \approx 2 \times 10^{24}$ m^{-3}, the pump power $P_p =$ 10 − 20 mW, the stimulated emission cross-section $\sigma = 8 \times 10^{-25}$ m^{-2}, the amplifier length $L =$ 10–20 m. The power gain can be as large as 10^4 (40 dB), with a conversion efficiency of 30–40 %.

The efficiency of the amplifier can be improved by using doubly-doped fibers containing both Er^{3+} and Yb^{3+} ions. In these fibers the pump power excites the ytterbium ions, which have an absorption cross-section larger than that of erbium ions. In a second step, there is a spontaneous transfer of excitation energy from ytterbium to erbium, so that a population inversion is built up in the erbium ions. This type of co-doping is mainly used for high-power amplifiers, with output power in the range of watts.

The main factor that limited the development of high-power amplifiers in early years was that the pump beam emitted by arrays of semiconductor lasers is spatially multimodal. The coupling efficiency of these multimodal beams inside the small core of the single-mode amplifying fiber is actually rather low. The problem is solved by using fibers with a double cladding, known as "cladding-pumped" fibers. The structure of these fibers is shown in Fig. 6.15. The signal to be amplified is confined within the core, whereas the pump is confined within the undoped inner cladding. Since the diameter of the inner cladding is much larger than that of the core, coupling of pump power into the fiber is much easier. The pump power is continuously fed from inner cladding to core, where it is absorbed, ensuring a high-efficiency operation. The efficiency of power transfer from inner cladding to core is improved by using an elliptical core.

In comparison with other optical-gain media, rare-earth doped-fibers have several very useful characteristics. The combination of broad gain bandwidth, polarization-independent gain, long lifetime of the excited level, and relatively large saturation

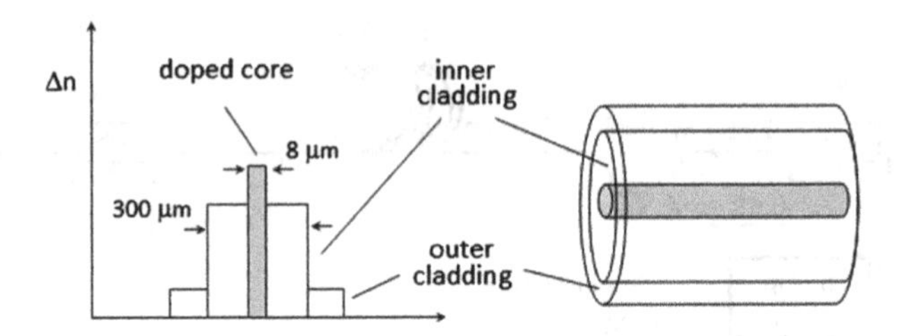

Fig. 6.15 Structure of a cladding-pumped optical fiber

power ensures full transparency to all modulation formats. The EDFA offers the possibility of simultaneous amplification of multiple signals at different wavelengths (like the multiple channels in a wavelength-division-multiplexing optical-communication system). There is also a low sensitivity of the EDFA gain to pump fluctuations occurring on a time-scale shorter than the lifetime of the excited level τ_{10}. Another important advantage is that the fiber geometry, with its large surface-to-volume ratio, has a good efficiency for heat dissipation.

Using schemes similar to those of EDFAs, fiber amplifiers operating at various near-infrared wavelengths can be made by doping silica fibers with Nd (amplification at 1.06 μ or 1.34 μm), or Yb (amplification in the range 1.05–1.12 μm). Optical amplifiers using Pr-doped ZBLAN fibers have also been tested for operation in the wavelength interval 1.28–1.34 μm, where some of the existing optical-communication systems operate.

A completely different family of fiber amplifiers is represented by Raman amplifiers, which make use of the stimulated Raman scattering process. They will be described in Chap. 7.

6.7 Fiber Lasers

It is clear from the general discussion of Sect. 3.3 that the fiber amplifier can be transformed into a laser by providing an optical feedback. The initial interest in fiber lasers came from the optical-communications area, because it was thought possible to use them as sources of high-repetition-rate pulse trains. Although the interest for such an application quickly vanished, fiber lasers have reached a high degree of reliability for use in many different areas.

Fiber lasers can operate both CW and pulsed. In comparison with solid-state lasers, they have several advantages.

- When using a single-mode fiber, the output beam has a very smooth Gaussian profile, therefore it can be focused to a very small focal spot, at variance with the behavior of high-power semiconductor or Nd-YAG lasers.
- The very good coupling of pump power into the fiber gives high-efficiency operation.

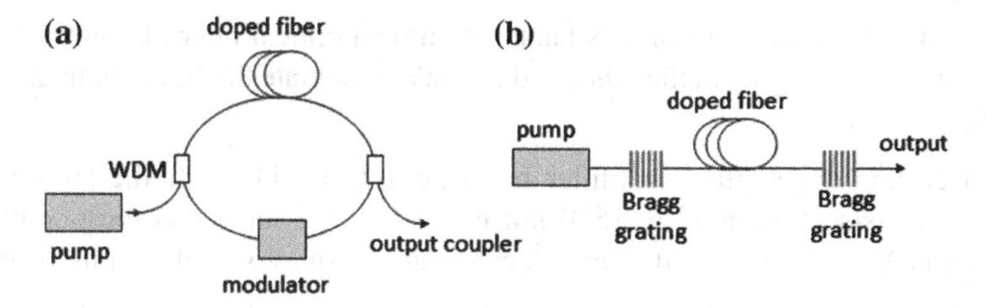

Fig. 6.16 Fiber laser schemes: **a** Ring cavity, **b** Fabry-Perot cavity

- The high surface-to-volume ratio of fiber structure provides an effective heat dissipation mechanism, so cooling problems are greatly simplified.

CW fiber lasers are based on Fabry-Perot cavities (see Fig. 6.16b) with mirrors made by fiber Bragg gratings. They use cladding pumped fibers in order to achieve high efficiency. A particularly interesting fiber laser is the ytterbium laser. The Yb^{3+} ion behaves like a four-level system, is pumped at 915 nm and emits in the 1000–1100 nm wavelength region. The emission wavelength is fixed by the peak reflectance of the FBG mirrors. The Yb-fiber laser can provide high-power outputs (up to some kilowatts). Because of its efficiency, compactness, and simplicity of cooling requirements, it is often used in industrial processes, such as marking, welding and precision cutting.

Pulsed fiber lasers can be built to operate in the Q-switching regime, generating kilohertz-frequency trains of nanosecond pulses, or in the mode-locking regime. Mode-locked fiber lasers are usually based on ring cavities (see Fig. 6.16a). The pump beam is injected into the doped fiber by using a dichroic fiber coupler. Another fiber coupler is then used to extract the output beam. The mode-locking behavior is induced by inserting a modulator or a saturable absorber inside the ring. These lasers can largely be based on telecom components, which have been developed for reliable long-term operation and have a moderate cost. Mode-locked erbium-doped or ytterbium-doped fiber lasers can generate trains of picosecond pulses with repetition rates up to 20 GHz. The very large gain bandwidth achievable in an ytterbium-doped fiber is capable of supporting pulses shorter than 20 fs.

Problems

6.1 A step-index fiber has a numerical aperture of 0.15, a core refractive index of 1.47 and a core diameter of 20 μm. Calculate (i) the maximum acceptance angle of the fiber, (ii) the refractive index of the cladding, (iii) the approximate number of modes with a wavelength of 800 nm that the fiber can carry.

6.2 An optical power of 10 mW is launched into an optical fiber of length 500 m. If the power emerging from the other end is 8 mW, calculate the fiber attenuation α_o in dB/km.

6.3 An erbium-doped fiber amplifier has length $L = 11$ m. If the stimulated-emission cross-section at $\lambda = 1530$ nm is 1.2×10^{-20} cm^2 and the population inversion is $N_2 - N_1 = 7 \times 10^{17}$ cm^{-3}, calculate the small-signal gain in dB units.

6.4 A light beam at $\lambda = 1550$ nm is coupled into a step-index optical fiber having numerical aperture $NA = 0.12$ and core radius $a = 4.5$ μm. (i) specify whether the fiber behavior will be monomodal or multimodal; (ii) calculate the parameter w of the fundamental mode.

6.5 A transform-limited Gaussian laser pulse having duration $\tau_p = 4$ ps and angular frequency ω_o is coupled into a single-mode optical fiber. The fiber has an attenuation coefficient $\alpha_o = 0.15$ dB/km and a frequency-dependent propagation constant $\beta(\omega)$ following the law:

$$\beta(\omega) = \beta_o + \beta_1(\omega - \omega_o) + (1/2)\beta_2(\omega - \omega_o)^2 \tag{6.39}$$

where $\beta_1 = 4.90 \times 10^{-9}$ s/m, $\beta_2 = -1 \times 10^{-27}$ s^2/m. By assuming that the fiber length is $L = 50$ km, calculate: (i) the time taken by the peak of the pulse to propagate along the fiber; (ii) the pulse duration $\tau_p(L)$ at the fiber output; (iii) the ratio r between the output and input energy of the laser pulse; (iv) the ratio $r' = P_L/P_0$, where P_0 and P_L are the input and output peak power of the laser pulse, respectively.

6.6 Consider a step-index optical fiber with a core radius $a = 4.5$ μm and core refractive index $n_1 = 1.494$. Assuming that the normalized frequency is $V = 2.15$ at $\lambda = 1550$ nm, calculate: (i) the cladding refractive index; (ii) the width w of the fundamental mode; (iii) the wavelength at which the fiber ceases to be single-mode.

6.7 An erbium-doped fiber amplifier has length $L = 10$ m and a stimulated emission cross section $\sigma = 1.2 \times 10^{-20}$ cm^2 at $\lambda = 1550$ nm. Calculate the value of the population inversion $N_2 - N_1$ required to attain a small signal gain of 20 dB.

6.8 Calculate pitch and length of a fiber Bragg grating having a reflectivity peak $R_{\max} = 0.995$ at $\lambda = 1500$ nm, assuming that the core average refractive index is $n = 1.5$ and that $\Delta n = 2 \times 10^{-4}$.

6.9 Calculate the effective refractive index n_{eff} at $\lambda = 1550$ nm for a single-mode optical fiber having a core radius $a = 5$ μm, core refractive index $n_1 = 1.484$, cladding refractive index $n_2 = 1.480$, by assuming that $n_{eff} = n_1 f_1 + n_2 f_2$, where f_1 and f_2 are the power fractions traveling inside the core and the cladding, respectively. The profile of the single mode is given by the Gaussian function $\exp[(x^2 + y^2)/w^2]$. The value of n_{eff} should be given with 4 digits.

6.10 A transform-limited Gaussian laser pulse having duration $\tau_p = 1$ ps is coupled into a single-mode optical fiber with a dispersion parameter $\beta_2 = 10$ ps^2/km. Calculate at which propagation distance the pulse duration is doubled.

Chapter 7
Nonlinear Optics

Abstract The field of nonlinear optics ranges from fundamental studies of the interaction of intense laser light with matter to applications such as frequency conversion, non-resonant amplification, all-optical switching. The treatment developed in this chapter restricts the attention to those topics that are of interest for applications, and limits the notational complications to a minimum. Second-harmonic generation is described in some detail in Sect. 7.2, because it is considered a case study for illustrating the mathematical approach and the experimental methods that are generally utilized in nonlinear optics. Section 7.3 completes the description of second-order effects with a concise treatment of parametric amplifiers and oscillators and of sum-frequency generation. Section 7.4 is devoted to a review of the main phenomena arising from the optical Kerr effect. The subject of Sect. 7.5 is stimulated Raman scattering, with particular emphasis on applications based on optical-fiber devices. Stimulated Brillouin scattering is described in Sect. 7.6.

7.1 Introduction

Nonlinear optics is the field of optics that considers phenomena occurring when the response of a medium to an applied electromagnetic field is nonlinear with respect to the amplitude of that field.

The proportionality relation between electric field and induced electric polarization that was introduced in Sect. 2.1 implies that a monochromatic field oscillating at ω generates a polarization P also oscillating at ω. As a consequence, the propagation of a wave across a transparent medium can change the spectrum by attenuating or suppressing some frequency component, but cannot generate new frequencies.

Nonlinear optical effects become appreciable when the displacement of charges from their equilibrium position inside the material becomes so large that the induced dipole ceases to be proportional to the field amplitude E. This happens if the incident field amplitude is not negligible in comparison with the internal field E_i that holds the charges together in the material. A typical value for E_i is 10^{11} V/m, a figure attainable with laser fields, whereas the electric-field amplitude of conventional optical sources can reach at most 10^3 V/m. Nonlinear optics was born in the early 1960s immediately after the first intense laser fields became available.

V. Degiorgio and I. Cristiani, *Photonics*, Undergraduate Lecture Notes in Physics,
DOI 10.1007/978-3-319-20627-1_7

The standard classification of nonlinear optical effects is based on the expansion of the electric polarization P in a series of powers of E:

$$P = \varepsilon_o(\chi^{(1)}E + \chi^{(2)}E^2 + \chi^{(3)}E^3 + \cdots). \tag{7.1}$$

Equation (7.1) can be considered as a generalization of the linear relation expressed by (2.11). For sake of simplicity, the notation adopted in (7.1) neglects dispersion and uses scalar quantities.

If propagation in a medium exhibiting a non-zero $\chi^{(2)}$ is considered, an input field oscillating at ω, $E(t) \propto \exp(-i\omega t)$, generates a polarization component $\propto \chi^{(2)} \exp(-2i\omega t)$, oscillating at 2ω. Once the expression of the electric polarization is inserted into the wave equation (2.7), the 2ω component becomes the source of a second-harmonic field. Analogously, if the input field contains two distinct frequencies ω_1 and ω_2, then it is found that the quadratic term generates the sum frequency $\omega_1+\omega_2$, and the difference frequency $\omega_1-\omega_2$. A similar discussion could be made for the cubic term. Besides optical frequency conversion, another feature of nonlinear interactions is that they can involve field-dependent modifications of certain material properties. This second feature opens the way to phenomena in which one beam can be used to control the amplitude and/or the phase of a second beam. To be more specific, through the optical Kerr effect a strong pump beam can change the index of refraction of the medium in which a signal beam propagates, so that a fully optical switch can be realized. In principle, nonlinear phenomena appear as the natural candidates for the realization of integrated all-optical devices, such as gates and logical circuits, to be used in various applications related to information and communication technology and optical signal processing.

7.2 Second-Harmonic Generation

Consider a monochromatic plane wave having angular frequency ω and wave vector $\mathbf{k}_\omega$ propagating along the y axis inside a second-order nonlinear medium characterized by a nonlinear susceptibility $\chi^{(2)}$. During propagation a second-harmonic wave will be generated, so that the total electric field of the waves traveling along y is:

$$\begin{aligned} E(y,t) &= \frac{1}{2}\{E_\omega(y)\exp[-i(\omega t - k_\omega y] + c.c.]\} \\ &\quad + \frac{1}{2}\{E_{2\omega}(y)\exp[-i(2\omega t - k_{2\omega} y] + c.c.]\}, \end{aligned} \tag{7.2}$$

where $k_\omega = (\omega/c)n_\omega$ and $k_{2\omega} = (2\omega/c)n_{2\omega}$. Here n_ω and $n_{2\omega}$ are, respectively, the refractive indices experienced by the input beam, called sometimes the pump beam, and by the second-harmonic beam during the propagation along the nonlinear medium.

Note that the field described by (7.2) is a real quantity, because of the presence of the complex conjugate terms. In the previous chapters many calculations have been carried out by representing the real electric field in terms of a complex function, whose real part is the field itself. Such an approach is justified only if the real field and its complex representation satisfy the same propagation equations. This is true as long as the equations are linear in the fields. The propagation equations treated in this chapter contain nonlinear terms, and so it is mandatory to deal with real fields.

The equations describing the y-dependence of the amplitudes E_ω and $E_{2\omega}$ are derived as follows. First, by neglecting the cubic term inside (7.1), and inserting (7.2) into (7.1), an expression for $P(y,t)$ is found. This expression will contain terms oscillating at ω, 2ω, 3ω, 4ω, and also time-independent terms. Considering only those terms that oscillate at ω and 2ω, $P(y,t)$ is given by:

$$P(y,t) = \frac{\varepsilon_o \chi^{(1)}}{2}[E_\omega e^{-i(\omega t - k_\omega y)} + c.c.] + \frac{\varepsilon_o \chi^{(1)}}{2}[E_{2\omega} e^{-i(2\omega t - k_{2\omega} y)} + c.c.] + \frac{\varepsilon_o \chi^{(2)}}{4}[E_\omega^2 e^{-2i(\omega t - k_\omega y)} + c.c.] + \frac{\varepsilon_o \chi^{(2)}}{4}[E_{2\omega} E_\omega^* e^{-i(\omega t - k_{2\omega} y + k_\omega y)} + c.c.] \tag{7.3}$$

As a second step, (7.2) and (7.3) are substituted inside the wave equation (2.7). By separately considering all the terms oscillating respectively as $\exp(-i\omega t)$ and $\exp(-2i\omega t)$, and further assuming that the terms containing the second-order derivative with respect to y can be neglected, the following two equations are obtained:

$$\frac{dE_\omega(y)}{dy} = i\frac{\omega\chi^{(2)}}{2cn_\omega} E_{2\omega} E_\omega^* \exp(i\Delta k y), \tag{7.4}$$

$$\frac{dE_{2\omega}(y)}{dy} = i\frac{\omega\chi^{(2)}}{2cn_{2\omega}} E_\omega^2 \exp(-i\Delta k y), \tag{7.5}$$

where $\Delta k = k_{2\omega} - 2k_\omega$. Taking the entrance face into the medium coincident with the plane y = 0, the initial conditions are:

$$E_\omega(0) \neq 0; \quad E_{2\omega}(0) = 0. \tag{7.6}$$

Recalling that the beam intensity is related to E_ω by:

$$I_\omega(y) = \frac{\varepsilon_o n_\omega c}{2}|E_\omega(y)|^2, \tag{7.7}$$

and noting that:

$$\frac{dI_\omega(y)}{dy} = \frac{\varepsilon_o n_\omega c}{2}\left(E_\omega \frac{dE_\omega^*}{dy} + E_\omega^* \frac{dE_\omega}{dy}\right), \tag{7.8}$$

the following relation is obtained from (7.4) and (7.5):

$$\frac{dI_\omega}{dy} + \frac{dI_{2\omega}}{dy} = 0 \tag{7.9}$$

Equation (7.9) indicates that, during propagation into the lossless nonlinear medium, there is a conservation of the overall optical power: the nonlinear medium acts only as a catalyst, and so the second-harmonic generation (SHG) process simply consists of an exchange of power between the two fields.

It is not possible to find an exact analytical solution to the system of (7.4) and (7.5). However, an approximate solution is immediately found by assuming that the amplitude of the generated second-harmonic wave is so weak that the amplitude of the fundamental wave is negligibly reduced by the SHG process. If $E_\omega(y)$ is considered a constant, coinciding with $E_\omega(0)$, then the solution of (7.5) is:

$$E_{2\omega}(y) = \frac{i\omega\chi^{(2)}}{cn_{2\omega}} E_\omega^2(0) \frac{1 - e^{-i\Delta k y}}{i\Delta k} \tag{7.10}$$

The second-harmonic (SH) intensity is given by:

$$I_{2\omega}(y) = I_\omega^2(0) y^2 \frac{2(\omega\chi^{(2)})^2}{c^2 n_{2\omega} n_\omega^2} \frac{4}{(\Delta k)^2} \sin^2\left(\frac{\Delta k y}{2}\right). \tag{7.11}$$

Equation (7.11) shows that $I_{2\omega}(y)$ is an oscillating function of y, at fixed Δk. This means that there is a continuous exchange of power between pump and SH beam during propagation. The first maximum of $I_{2\omega}(y)$ occurs at the propagation distance $L_c = \pi/\Delta k$, known as the coherence length of second-harmonic generation. If the propagation distance y exceeds L_c, the SH power starts to decrease because the newly generated SH contributions have an opposite phase with respect to those generated for $y < L_c$. It is important to note that the amplitude of the spatial oscillation gets bigger and bigger as Δk decreases. Of course, if the SH power becomes large, the approximation of no-pump-depletion is no longer applicable.

The conversion efficiency, $\eta_{2\omega}$, defined as the ratio between SH intensity at the output and pump intensity at the input, is expressed as:

$$\eta_{2\omega} = \frac{I_{2\omega}(L)}{I_\omega(0)} = I_\omega(0) L^2 \frac{2(\omega\chi^{(2)})^2}{c^2 n_{2\omega} n_\omega^2} \mathrm{sinc}^2\left(\frac{\Delta k L}{2}\right), \tag{7.12}$$

where L is the propagation length, and $\mathrm{sinc}(x) = \sin x/x$. The efficiency $\eta_{2\omega}$ is maximum if

$$\Delta k = k_{2\omega} - 2k_\omega = 0, \tag{7.13}$$

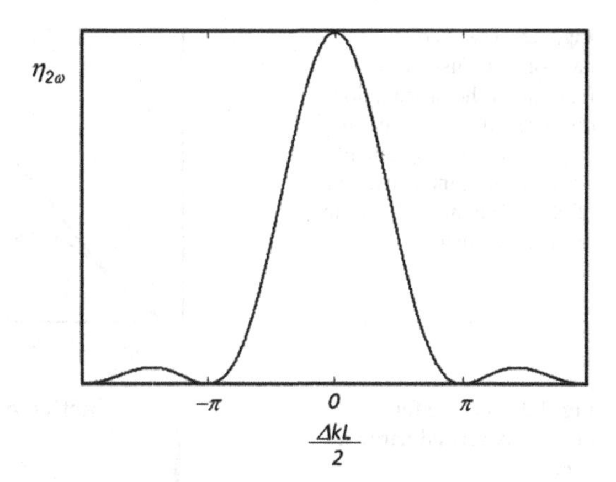

Fig. 7.1 Efficiency of the second-harmonic generation process plotted as a function of the phase mismatch

as shown in Fig. 7.1. This condition, known as the "phase-matching" condition, implies that

$$n_{2\omega} = n_\omega, \tag{7.14}$$

i.e. the SH wave should travel at the same phase velocity as the pump wave, so that all the contributions generated during propagation can add in phase.

In general terms, time-invariance and translational invariance (which means monochromatic plane waves interacting in an infinitely extended crystal) impose energy and momentum conservation among the interacting waves. Using the photon concept, energy conservation is satisfied when two pump photons are converted into one SH photon, and momentum conservation requires that the momentum of the generated SH photon is the sum of the momenta of the two annihilated pump photons. Therefore the phase-matching condition can be seen as a momentum conservation condition.

By assuming $\Delta k = 0$, (7.4) and (7.5) can be exactly solved, obtaining:

$$I_{2\omega}(L) = I_\omega(0)\tanh^2(L/L_{SH}), \tag{7.15}$$

and

$$I_\omega(L) = I_\omega(0)\text{sech}^2(L/L_{SH}), \tag{7.16}$$

where the length L_{SH} is defined as

$$L_{SH} = \frac{\lambda\sqrt{n_\omega n_{2\omega}}}{\pi\chi^{(2)}E_\omega(0)}. \tag{7.17}$$

It is easy to verify that $I_{2\omega}(L) + I_\omega(L) = I_\omega(0)$ for any L. Recalling that the asymptotic limit of the hyperbolic tangent for L going to infinity is equal to 1,

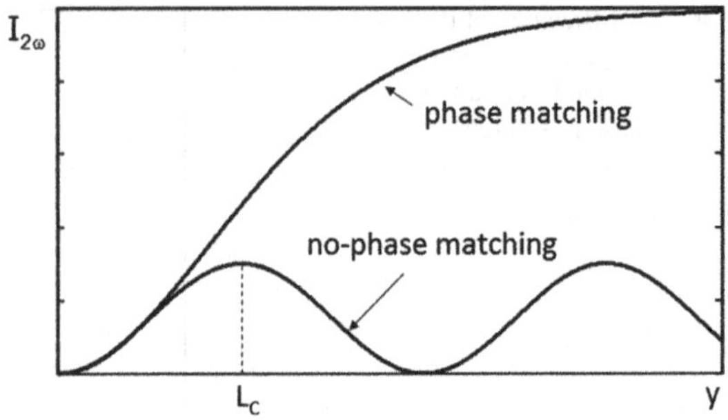

Fig. 7.2 The second-harmonic intensity as a function of the propagation distance. At phase matching $I_{2\omega}$ asymptotically coincides with the input intensity. Out of phase matching, $I_{2\omega}$ is an oscillatory function

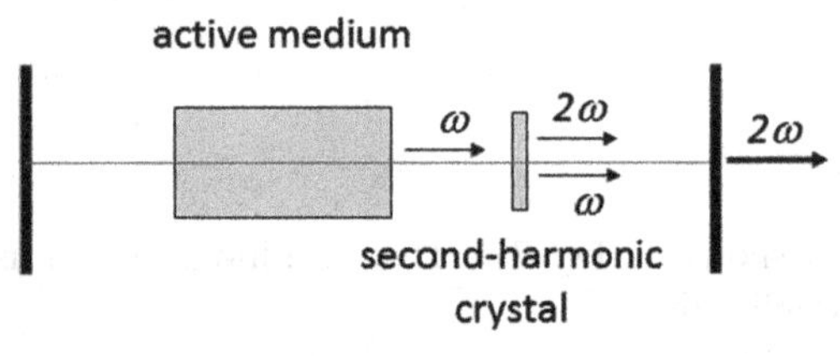

Fig. 7.3 Scheme for intracavity second-harmonic generation

(7.15) shows that the input intensity can be fully converted into the second-harmonic intensity, provided that the propagation distance is sufficiently long (the crystal length L should be much larger than L_{SH}). The behavior of the SH intensity as a function of the propagation distance is illustrated in Fig. 7.2.

The first nonlinear optical experiment was performed in 1961 by Franken and coworkers by using quartz as the nonlinear crystal. The pump was a ruby laser pulse at $\lambda = 0.6943\ \mu$m. A very weak second-harmonic signal at $\lambda = 0.347\ \mu$m was detected. In that experiment the conversion efficiency was very low, about 10^{-8}, because the phase-matching condition was not satisfied.

SHG becomes particularly efficient if the nonlinear crystal is placed inside the laser cavity, as schematically shown in Fig. 7.3. Intra-cavity SHG represents a very convenient method to generate coherent green and blue light starting from near infrared solid-state lasers. As an example, 532-nm green light can be generated by frequency-doubling of a neodymium laser in a Fabry-Perot cavity in which both mirrors are totally reflecting at 1064 nm and the output mirror is totally transmitting at 532 nm. Once fixed the gain in the active medium, the type of nonlinear crystal, and the cavity structure, the only free design parameter is the length L of the crystal. This is a typical optimum-coupling problem: if L is too small there is little second-harmonic generation, if L is too large there are strong cavity losses.

7.2.1 Second-Order Nonlinear Optical Materials

Similarly to the case of the Pockels effect described in Chap. 4, second-order non-linear optical interactions can occur only in non-centrosymmetric materials, that is, in materials that do not display inversion symmetry. Since liquids, gases, amorphous solids (such as glass), and even many crystals do, in fact, display inversion symmetry, $\chi^{(2)}$ vanishes identically for such media, and consequently they cannot produce second-order nonlinear optical interactions. Second-order optical materials are intrinsically anisotropic, thus electric susceptibilities of any order are tensorial quantities. The i-th cartesian component of the vector **P** is expressed as:

$$P_i = \varepsilon_o\Big(\sum_{j=1}^{3} \chi_{ij} E_j + \sum_{j,k=1}^{3} \chi_{ijk} E_j E_k\Big) \tag{7.18}$$

Equation (7.18) represents an extension of the linear treatment developed in Sect. 2.6 for optically anisotropic media. Assuming that the second-order tensor χ_{ij} is diagonalized, the only non-zero components of the linear susceptibility are χ_{11}, χ_{22}, and χ_{33}. Furthermore, in order to simplify the notation, only crystals possessing a symmetry axis, taken as the z-axis, will be considered. In such crystals, $\chi_{11} = \chi_{22}$.

Note that (7.18) has the same structure as (4.2) that was written to describe the Pockels effect, and so the properties of the third-order tensor $\chi^{(2)}$ are similar to those of the electro-optic tensor, with an important difference concerning the symmetry properties. Since the value of P_i cannot change when we exchange the position of E_k with that of E_j inside (7.18), the symmetry relation $\chi_{ijk} = \chi_{ikj}$ holds in this case. In the scientific and technical literature, the properties of second-order nonlinear materials are usually expressed in terms of the d_{ijk} tensor that is defined by:

$$d_{ijk} = \frac{1}{2}\chi_{ijk} \tag{7.19}$$

In the standard notation only two indices are used, by replacing the pair jk with one index m running from 1 to 6. The following convention is used: $m = 1$ corresponds to $jk = 11$, $m = 2$ to $jk = 22$, $m = 3$ to $jk = 33$, $m = 4$ to $jk = 23$, $m = 5$ to $jk = 13$, and $m = 6$ to $jk = 12$. The nonlinear susceptibility tensor d_{im} can then be represented as a 3×6 matrix.

Crystalline structures are classified according to 32 distinct symmetry classes. For each class it is possible to predict, using symmetry arguments, what tensor components are non-zero. For instance, in the case of lithium niobate the non-zero components are: d_{15}, d_{16}, d_{21}, d_{22}, d_{24}, d_{31}, d_{32}, and d_{33}.

In order to illustrate the basic aspects of nonlinear interactions, it is useful to discuss some specific situations. Consider a pump beam propagating along the y axis of a lithium niobate crystal. If the beam is an ordinary wave, linearly polarized along x, the index m can only take the value 1, so that, among all the non-zero components of d_{im}, the only accessible ones are d_{21} and d_{31}. Since the propagation is along y, the electric polarization cannot have a longitudinal component, so d_{21} does not play any role in the SHG process. The only surviving terms from the general relation (7.18) are:

Table 7.1 Symmetry class, transparency window and nonlinear coefficients for some nonlinear crystals at $\lambda = 1064$ nm

Crystal	Class	Window (μm)	d (10^{-12} m/V)
$LiNbO_3$	3 m	0.35–5.20	$d_{33} = 25.2,\ d_{31} = 4.6$
			$d_{22} = 2.1$
KTP	mm^2	0.35–4.50	$d_{33} = 13.7,\ d_{31} = 6.5$
			$d_{32} = 5.0$
LBO	mm^2	0.16–2.30	$d_{33} = 0.04,\ d_{31} = 0.85$
			$d_{32} = 0.67$

Lithium niobate ($LiNbO_3$), potassium titanyl phosphate ($KTiOPO_4$), known as KTP, lithium triborate (LiB_3O_5), known as LBO

$$P_1(\omega) = \varepsilon_o \chi_{11} E_1(\omega)$$
$$P_3(2\omega) = 2\varepsilon_o d_{31}[E_1(\omega)]^2. \tag{7.20}$$

The component $P_3(2\omega)$ generates a SH field polarized along z, i.e. an extraordinary SH wave. Therefore the phase-matching condition becomes:

$$\Delta k = \frac{2\omega}{c}(n^e_{2\omega} - n^o_{\omega}) = 0. \tag{7.21}$$

If the pump beam is linearly polarized along z, the only surviving terms from the general relation (7.18) are:

$$P_3(\omega) = \varepsilon_o \chi_{33} E_3(\omega)$$
$$P_3(2\omega) = 2\varepsilon_o d_{33}[E_3(\omega)]^2 \tag{7.22}$$

In this case, both pump and SH fields are extraordinary waves, and the phase-matching condition is:

$$\Delta k = \frac{2\omega}{c}(n^e_{2\omega} - n^e_{\omega}) = 0. \tag{7.23}$$

In order to prepare the description of the different phase-matching methods that will be given in the next section, it is instructive to examine the properties of some common nonlinear crystals—see Table 7.1.

Lithium niobate and KTP exhibit large nonlinear coefficients. LBO has smaller coefficients, but its transparency window extends into the ultraviolet, and thus is the best candidate to generate ultraviolet light by SHG.

7.2.2 Phase Matching

The practical importance of nonlinear optics comes from the fact that the new fields generated by nonlinear interactions can take up, in principle, all the optical power of the incoming fields. However, as shown in the preceding section, this is possible only if phase-matching is satisfied. Phase matching should be seen not only as a constraint that limits the applicability of nonlinear effects, but also as an effective method to select a single nonlinear interaction, from among many possible interactions. The structure of the equations is greatly simplified if one defines a priori which interaction will be phase-matched. According to this criterion, only those terms relevant for SHG have been selected in writing (7.3).

Birefringence phase matching. Since the index of refraction of any medium changes with ω, it is generally not possible to satisfy the phase-matching condition (7.23). In contrast, condition (7.21) suggests the possibility of obtaining phase matching by exploiting the crystal birefringence to compensate for material dispersion (see Fig. 7.4). As an example, lithium niobate is a uniaxial crystal presenting negative birefringence ($n_o > n_e$), as indicated in Table 2.3. Taking into account that both indices grow with ω, it can be shown that there is a frequency ω_{pm} (corresponding to a wavelength $\lambda_{pm} \approx 1.1\ \mu\text{m}$), for which: $n_o(\omega_{pm}) = n_e(2\omega_{pm})$. In such a situation, an ordinary input beam can be efficiently converted into an extraordinarily polarized second-harmonic beam.

For a given crystal the condition $n_o(\omega_{pm}) = n_e(2\omega_{pm})$ can be at most satisfied for only one wavelength. There are, however, some degrees of freedom that can be exploited to obtain tunable phase matching. One possibility is offered by the consideration that the refractive index of the extraordinary beam can be varied continuously between n_o and n_e by changing the angle between $\mathbf{k}_\omega$ and the optical axis, as described by (2.128). Therefore, one can tune ω_{pm} by rotating the crystal around the axis perpendicular to the plane containing $\mathbf{k}_\omega$ and the z axis. In this case there is, however, a non-zero spatial walk-off of the second-harmonic beam with respect to the fundamental beam, because the two Poynting vectors are no longer parallel (see

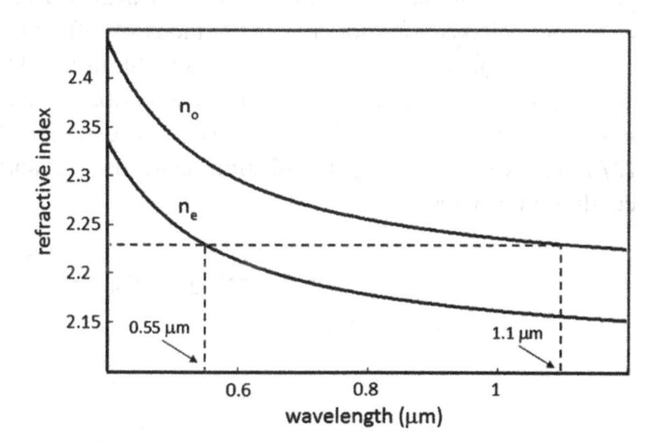

Fig. 7.4 Birefringence phase matching in lithium niobate

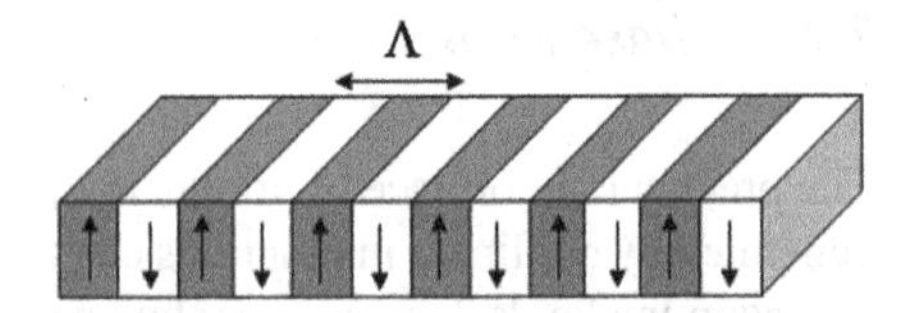

Fig. 7.5 Nonlinear crystal with periodic inversion of the ferroelectric domains

(2.129) in Chap. 2). Since the spatial overlap between the two beams decreases with the propagation distance, the walk-off effect introduces a limitation of the crystal length. Noting that the refractive indices are weakly dependent on temperature, it is also possible to finely tune ω_{pm} by varying T.

The method so far discussed, in which two ordinary photons at frequency ω are converted into one extraordinary photon at frequency 2ω is known as type I phase matching. Other degrees of freedom are added to the SH experiment if two pump beams having the same frequency, but different directions or different polarization, are considered. A particularly interesting method, known as type II phase matching, uses one ordinary and one extraordinary photon to generate the SH photon. The phase-matching condition for this method is:

$$n_{\omega}^{o} + n_{\omega}^{e} = 2n_{2\omega}^{e}. \tag{7.24}$$

Quasi-phase-matching. There are circumstances in which birefringence phase-matching is not suitable for an efficient SHG. For instance, the nonlinear material may possess insufficient birefringence to compensate for linear dispersion. Another case is when the application requires the use of the d_{33} coefficient, which is much larger than the off-diagonal coefficients for some crystals (see Table 7.1). The d_{33} coefficient can be accessed only if all the interacting waves have the same polarization, but the problem is that the phase-matching condition (7.24) cannot be satisfied because of optical dispersion. There is a method, known as quasi-phase-matching (QPM), that is alternative to birefringence phase-matching. This method makes use of a periodically poled nonlinear crystal. As shown in Fig. 7.5, such a material consists of a sequence of slabs parallel to the optical axis, each having a thickness $\Lambda/2$ and presenting an alternate orientation of the optical axis. The inversion of the orientation of the optical axis does not change the index of refraction, but has the consequence of inverting the sign of the nonlinear susceptibility, so that now d_{33} is not constant along y, but exhibits a square spatial oscillation with period Λ. Considering the first harmonic of the function $d_{33}(y)$, which is expressed by the sinusoidal function $(2/\pi)d_{33}\exp[i(2\pi/\Lambda)y]$, and substituting inside (7.4) and (7.5), the phase-matching condition becomes:

$$\Delta k = \frac{2\omega}{c}(n_{2\omega}^{e} - n_{\omega}^{e}) = \frac{2\pi}{\Lambda}. \tag{7.25}$$

Equation (7.25) implies that $\Lambda = 2L_c$. Note that, in the case of QPM, the momentum of the SH photon is not equal to twice the momentum of the pump photon. The difference is taken up by the periodically poled crystal.

An intuitive explanation of the QPM can be given as follows. In the presence of a wavevector mismatch, the amplitude of the generated SH wave oscillates with propagation distance, as shown in Fig. 7.2. If the period Λ of the alternation of the optical axis is set equal to twice L_c, then each time the SH amplitude is about to begin to decrease because of the wavevector mismatch, a reversal of the sign of d_{33} occurs which allows the SH amplitude to grow monotonically.

Equation (7.25) indicates that phase matching can be achieved at any desired frequency by appropriately choosing Λ. An important property of the periodical-poling approach to phase-matching is also that it is effective with any nonlinear crystal, irrespective of the fact that the material might be birefringent or not. The technical problem is how to fabricate a periodically poled crystal. If the nonlinear optical material is a ferroelectric crystal, as is the case for lithium niobate and KTP, the orientation of the optical axis can be defined by a sufficiently strong static electric field. Therefore, periodically poled structures can be fabricated by applying electric fields with a sequence of electrodes of alternating polarity. Typically, electric fields of the order of 20 kV/mm are applied, with a period that can be as small as 5 μm.

As an example, consider a second-harmonic-generation experiment using a neodymium laser ($\lambda = 1064$ nm) and a KTP crystal. By putting $2d_{31}$ instead of $\chi^{(2)}$ inside (7.17), and taking $d_{31} = 6.5$ pm/V, it is found that, in order to obtain a reasonable value for L_{SH}, say 1 cm, it is necessary to have an input field $E_\omega(0) \approx 4.7$ MV/m, corresponding to a laser intensity of about 5 MW/cm^2. This can be obtained by focusing the laser output or by coupling the laser beam into a nonlinear single-mode waveguide. The required intensity is however reduced by a factor 16, if QPM is used to exploit the larger coefficient of d_{33} instead of d_{31}.

7.2.3 SHG with Ultrashort Pulses

Spectral acceptance. The conversion efficiency $\eta_{2\omega}$, defined by (7.12), reduces to half of the peak value when $\Delta kL \approx \pi/2$, as shown in Fig. 7.1. One can therefore define a phase-matching bandwidth $\Delta k_{pm} \approx \pi/L$, and, correspondingly, a frequency bandwidth $\Delta\omega_{pm}$, usually known as the spectral acceptance. A pump beam at an angular frequency ω can efficiently generate a SH beam provided that $\omega_{pm} - \Delta\omega_{pm}/2 \leq \omega \leq \omega_{pm} + \Delta\omega_{pm}/2$. The expression for $\Delta\omega_{pm}$ can be derived by inserting the first-order frequency-dependence of the refractive indices into the relation

$$\Delta k = (2\omega/c)n^e_{2\omega} - (2\omega/c)n^o_{2\omega} = \pi/L,$$

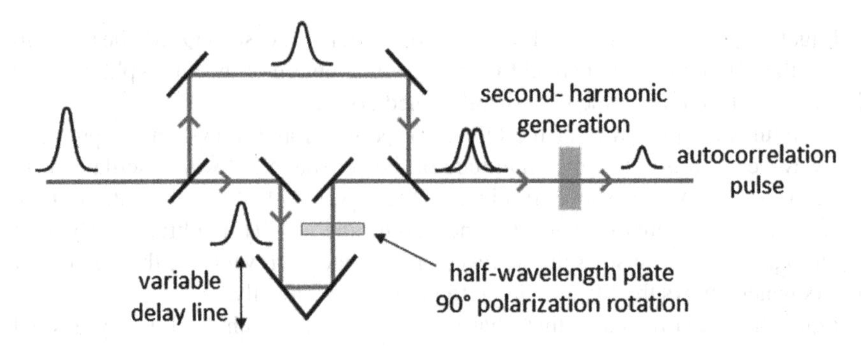

Fig. 7.6 Optical scheme of an autocorrelator for the measurement of duration of ultrashort pulses

and obtaining:

$$\Delta\omega_{pm} = \frac{\lambda_{pm}}{L}\left(\frac{dn^o_\omega}{d\omega} - 2\frac{dn^e_{2\omega}}{d\omega}\right)^{-1}. \tag{7.26}$$

The derivatives in (7.26) are calculated at $\omega = \omega_{pm}$.

The concept of spectral acceptance is very useful when dealing with nonlinear interactions involving optical pulses. As shown in in Sect. 3.11.3, a transform-limited pulse having a duration τ_p has a spectral broadening $\Delta\omega_p$ of the order of $1/\tau_p$. All the optical frequencies present in the pulse can contribute to SHG only if $\Delta\omega_p$ is smaller than $\Delta\omega_{pm}$. Considering that $\Delta\omega_{pm}$ is inversely proportional to L, the shorter is the pulse, the smaller the crystal thickness should be.

Measurement of the duration of ultrashort pulses. The optical pulses generated by mode-locked lasers usually have a duration much shorter than the response time of the photodetectors (see Sect. 5.5), so it is not possible to measure their duration by direct detection. The measurement can be performed through second-harmonic generation by the method illustrated in Fig. 7.6. It involves dividing the pulse into two paths using a beam-splitter, then introducing a variable delay between the pulses, and then spatially overlapping the two pulses in an instantaneously responding nonlinear-optical medium, such as a SHG crystal. The best performance is obtained using the type-*II* phase matching described by (7.24). The linear polarization of one of the two pulses is rotated, so that, when they recombine at the SHG crystal, one acts as an ordinary wave and the other as an extraordinary wave. Type-*II* phase-matching requires that second-harmonic photons are exclusively generated by using one photon from each pulse, therefore the SH signal appears only if there is a temporal overlap between the two pulses. The detector response time is much longer than the pulse duration, so that the detected SH signal is an integral over time t. By sweeping the delay τ between the two pulses, the measured SH power versus τ becomes proportional to the autocorrelation function of the pulse intensity $I(t)$:

$$G(\tau) = \int_{-\infty}^{\infty} I(t)I(t+\tau)dt. \tag{7.27}$$

In the case of a Gaussian pulse of duration τ_p, $G(\tau)$ is also Gaussian, with a width at half-height equal to $\sqrt{2}\tau_p$.

Since the delay is changed by changing the path length, and path length variations can be controlled with an accuracy of the order of one micron, the time resolution of the measurement is of the order of 1 fs.

7.3 Parametric Effects

Difference-frequency generation. If two beams at frequencies ω_1 and ω_3, with $\omega_3 > \omega_1$, are sent into a second-order nonlinear crystal, then the nonlinear interactions can give rise to new frequencies representing either the sum or the difference between the input frequencies. By choosing a specific phase-matching condition, only one process can be selected. In the case of difference-frequency generation (DFG), a new beam at frequency $\omega_2 = \omega_3 - \omega_1$ is generated, provided that the condition $\mathbf{k_2} = \mathbf{k_3} - \mathbf{k_1}$ is satisfied. All the phase-matching methods discussed in the previous section in connection with SHG can also be applied to sum/difference-frequency generation. If birefringence phase-matching is used, then each of the three waves can be ordinary (o) or extraordinary (e). For instance, type I phase matching in a negative birefringence crystal, like LBO or KTP, is described as $o - o - e$, indicating that waves at ω_1 and ω_2 are ordinary, and the highest frequency ω_3 beam is extraordinary.

In the DFG process, one photon at frequency ω_3 is annihilated and two photons at frequencies ω_1 and ω_2 are created, so that the process not only produces a new coherent beam at ω_2, but also amplifies the beam at ω_1. Such an amplification process is known as "parametric amplification". Whereas standard optical amplifiers work only at frequencies corresponding to transitions among specific energy levels, parametric amplifiers use transparent crystals and can amplify, in principle, any frequency within the transparency window of the crystal, provided that phase matching can be achieved.

Considering collinear propagation along the y axis, the total field is given by the superposition of three fields:

$$E(y,t) = \frac{1}{2}\{E_1(y)\exp[-i(\omega_1 t - k_1 y)] + c.c. + E_2(y)\exp[-i(\omega_2 t - k_2 y)] + c.c. + E_3(y)\exp[-i(\omega_3 t - k_3 y)] + c.c.\} \quad (7.28)$$

The nonlinear propagation equations, derived by using the same approach outlined for SH generation, are:

$$\frac{dE_1(y)}{dy} = i\frac{\omega_1\chi^{(2)}}{2cn_1}E_3E_2^*\exp(i\Delta ky), \quad (7.29)$$

$$\frac{dE_2(y)}{dy} = i\frac{\omega_2\chi^{(2)}}{2cn_2}E_3E_1^* \exp(i\Delta ky), \tag{7.30}$$

$$\frac{dE_3(y)}{dy} = i\frac{\omega_3\chi^{(2)}}{2cn_3}E_1E_2 \exp(-i\Delta ky), \tag{7.31}$$

where $\Delta k = k_3 - k_1 - k_2$.

The situation here considered is that of a high intensity ω_3 beam, called pump, and a low intensity ω_1 beam, called signal. The generated ω_2 beam is called idler. Equations (7.29) and (7.30) can be analytically solved at phase matching by assuming that the pump beam is not depleted by the DFG process, such that $E_3(y)$ is considered as a constant, coinciding with $E_3(0)$. Fixing the initial conditions as $E_1(0) \neq 0$ and $E_2(0) = 0$, the obtained solutions are:

$$\begin{aligned} E_1(y) &= E_1(0)\cosh(g_p y) \\ E_2(y) &= i\sqrt{\frac{n_1\omega_2}{n_2\omega_1}}E_1(0)\sinh(g_p y), \end{aligned} \tag{7.32}$$

where the parametric gain per unit length g_p is given by:

$$g_p = \frac{1}{2c}\sqrt{\frac{\omega_1\omega_2}{n_1n_2}}\chi^{(2)}E_3(0). \tag{7.33}$$

Equation (7.32) shows that both fields grow monotonically as functions of the propagation distance. Since the annihilation of a photon at ω_3 simultaneously produces a photon at ω_1 and a photon at ω_2, the two intensities I_1 and I_2 grow in a parallel way: $I_1(y) = (\omega_1/\omega_2)I_2(y) + I_1(0)$. As an example, consider the case of a lithium niobate crystal pumped by a 532-nm beam having an intensity of 5 MW/cm^2, producing a field $E_3(0) = 4.1 \times 10^6$ V/m. Taking $\chi^{(2)} = 10$ pm/V, $n_1 = n_2 = 2.2$, $\nu_1 = 3 \times 10^{14}$ Hz, it is found: $g_p = 0.6$ cm^{-1}.

Recalling that an amplifier becomes an oscillator once a positive feedback is provided, the parametric amplifier can be transformed into a parametric oscillator by using an optical cavity (see Fig. 7.7). The initial seed for parametric oscillation comes from parametric fluorescence, that is the process by which a "pump" photon at frequency ω_3 spontaneously breaks down into two photons at frequencies ω_1 (signal) and ω_2 (idler). Potentially, parametric fluorescence can yield any pair of frequencies that satisfy energy conservation, but the choice must also satisfy the phase-matching condition. Tuning of the parametric oscillator is usually achieved by rotating the crystal or, in a finer scale, by changing the operating temperature.

Sum-frequency generation. Considering a situation in which the phase-matching condition is satisfied for the sum-frequency generation (SFG), and assuming again that the pump at ω_3 is not depleted, the two equations describing the evolution of the signal field at ω_1 and the sum field at $\omega_2 = \omega_1 + \omega_3$ are the following:

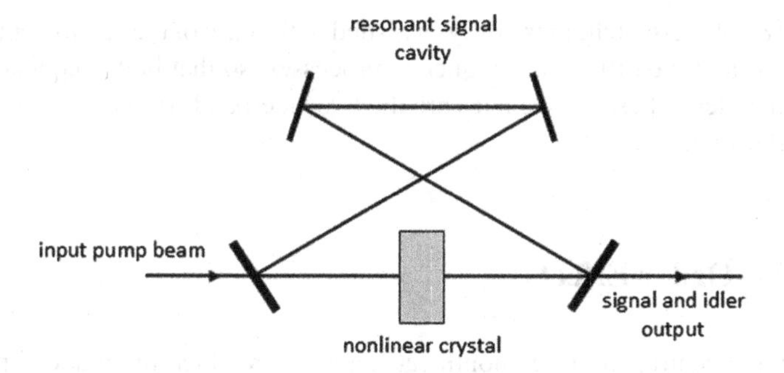

Fig. 7.7 Oscillator cavity based on parametric amplification

$$\frac{dE_1(y)}{dy} = i\frac{\omega_1\chi^{(2)}}{2cn_1}E_3^*(0)E_2$$
$$\frac{dE_2(y)}{dy} = i\frac{\omega_2\chi^{(2)}}{2cn_2}E_3(0)E_1. \tag{7.34}$$

Without loss of generality, one can put $E_3(0) = E_3^*(0)$, and the solutions are:

$$E_1(y) = E_1(0)\cos(g'_p y)$$
$$E_2(y) = -i\sqrt{\frac{n_1\omega_2}{n_2\omega_1}}E_1(0)\sin(g'_p y), \tag{7.35}$$

where

$$g'_p = \frac{1}{2c}\sqrt{\frac{\omega_1\omega_2}{n_1n_2}}\chi^{(2)}E_3(0). \tag{7.36}$$

The intensity conversion ratio in a crystal of length L is:

$$\frac{I_2(L)}{I_1(0)} = \frac{\omega_2}{\omega_1}\sin^2(g'_p L). \tag{7.37}$$

In the case in which all the input ω_1 photons are converted into ω_2 photons, the conversion ratio has the maximum value ω_2/ω_1. This is larger than one, reflecting the fact that the intensity of the converted beam is the sum of $I_1(0)$ plus the intensity $(\omega_3/\omega_1)I_1(0)$ provided by the pump beam.

SFG can be useful for converting an infrared beam into a more easily detectable visible beam, or for generating ultraviolet light by mixing visible and infrared light.

As a final comment to this section, it should be noted that the linear electro-optic effect (Pockels effect) of Sect. 4.1 can be considered as a particular case of frequency mixing, in which the external electric field $\mathbf{E}'$ at frequency ω_m plays the role of the field $\mathbf{E}_1$. Since ω_m is negligibly small with respect to the optical frequency ω, phase-matching is always satisfied in the case of the electro-optic effect for the sum- and the difference-frequency processes, so that both frequencies $\omega \pm \omega_m$ are generated. These frequencies are the two side-bands of the electro-optically modulated signal.

7.4 Third-Order Effects

Considering a centrosymmetric nonlinear material, in which the induced polarization is the sum of a linear term plus a cubic term, and assuming that the material is illuminated by a monochromatic plane wave having angular frequency ω and propagating along the y-axis, the electric polarization inside the nonlinear medium can be written as:

$$P(y,t) = \frac{\varepsilon_o \chi^{(1)}}{2}\left[E_\omega e^{-i(\omega t - k_\omega y)} + c.c.\right] + \frac{\varepsilon_o \chi^{(3)}}{8}\left[E_\omega^3 e^{-3i(\omega t - k_\omega y)} + c.c.\right] + \frac{3\varepsilon_o \chi^{(3)}}{8}\left[E_\omega^2 E_\omega^* e^{-i(\omega t - k_\omega y)} + c.c.\right]. \quad (7.38)$$

Many different nonlinear phenomena can be generated by the third-order nonlinearity. Here the description will be limited to phenomena occurring when a single beam at a single frequency and direction is incident on the nonlinear medium. Wave mixing processes involving more than one incident frequency will not be discussed.

Third-harmonic generation. The electric polarization of (7.38) contains terms oscillating at ω and at 3ω. Once the expression for $P(y,t)$ is inserted into the wave equation, the 3ω term will give rise to third-harmonic generation (THG). Formally, one could expect a behavior of the THG conversion efficiency similar to that found for SHG. However, the practical use of THG is rather limited because it is difficult to find materials simultaneously having a large $\chi^{(3)}$, a transparency window including both ω and 3ω, and also offering the possibility of phase matching. In practice, THG is usually obtained through a cascade of two second-order processes.

Optical Kerr effect. Neglecting the 3ω term, (7.38) can be re-written as $P(y,t) = P_\omega \exp[-i(\omega t - k_\omega y)] + c.c.$, where:

$$P_\omega = \varepsilon_o\left[\chi^{(1)} + \frac{3\chi^{(3)}}{4}|E_\omega|^2\right]E_\omega = \varepsilon_o \chi_{eff}^{(1)} E_\omega, \quad (7.39)$$

Equation (7.39) is similar to (4.37), which describes the Kerr effect. The big difference is that, in this case, the change in the susceptibility of the medium is due to the electric field of the propagating wave, whereas, in the case treated in Sect. 4.2, the effect was induced by a "low-frequency" external field created by an electronic device.

By using the relation between index of refraction and electric susceptibility, and assuming that the nonlinear contribution is small in comparison with the linear one, the effective refractive index is:

$$n_{eff} = \sqrt{1 + \chi_{eff}^{(1)}} \approx n + n_2' |E_\omega|^2 \quad \text{or} \quad n_{eff} = n + n_2 I, \tag{7.40}$$

where I is the intensity of the light beam and the nonlinear coefficient n_2, measured in m^2/W, is given by:

$$n_2 = \frac{3\chi^{(3)}}{4\varepsilon_o c n^2}. \tag{7.41}$$

Equation (7.40) expresses the optical Kerr effect. Values of n_2 are given in Table 7.2 for three materials.

As already discussed for the Kerr effect in Sect. 4.2, a more detailed analysis would show that an initially isotropic material becomes uniaxial in the presence of a strong linearly polarized optical beam, with the direction of the optical axis coinciding with the direction of the incident electric field. The index ellipsoid of the isotropic medium, initially spherical, acquires an intensity-dependent deformation. If the optical Kerr effect is only due to the displacement of electron clouds, the induced birefringence almost instantaneously follows changes in the intensity of the illuminating beam. Instead of considering a self-induced effect, one could imagine a cross-induced situation, in which a high-intensity beam modulates a weak probe beam through the optical Kerr effect, with a response time as short as the duration of the pump pulse (less than 1 ps). By using the same schemes of Sect. 4.1, a very fast amplitude- or phase-modulation could be achieved. In principle, this opens the way to the realization of all-optical modulators and switches.

It is interesting to discuss a numerical example involving the common material of silica glass, which has $\chi^{(3)} = 1.8 \times 10^{-22}$ m^2/V^2, corresponding to $n_2 = 2.4 \times 10^{-20}$ m^2/W, as reported in Table 7.2. A propagation along the distance L inside the glass produces a nonlinear phase delay $\Delta\phi = (2\pi/\lambda) n_2 I L$. Taking $L = 1$ cm and $\lambda = 1$ μm, the light intensity required to induce a nonlinear phase shift equal to π is: $I = 2.1 \times 10^{15}$ W/m^2. Such an intensity could be reached by focusing a laser pulse with energy of 2 μJ and a duration of 1 ps into a focal spot of diameter 30 μm. Such an optical switch is conceptually interesting from the point

Table 7.2 Two-photon absorption and Kerr coefficients for silica glass, silicon (Si) and gallium arsenide (GaAs) at two distinct wavelengths

Material	λ (nm)	β_a (cm/GW)	n_2 (cm^2/W)
Silica glass	1064		2.4×10^{-16}
Si	1540	0.8	0.45×10^{-13}
GaAs	1540	10.2	1.59×10^{-13}

of view of realizing all-optical logical circuits, but has little practical applicability, because it is bulky and requires too large a pulse energy. However, the performance can be greatly improved by using a material with a larger n_2 and confining the optical beams inside a waveguide of small cross-section.

Two-photon absorption. In general terms, the third-order susceptibility can be a complex quantity, written as: $\chi^{(3)} = \chi_{Re}^{(3)} + i\chi_{Im}^{(3)}$. In order to see the effect of the imaginary part, consider a plane wave, with electric field $(1/2)E(y)\exp[-i(\omega t - ky)] + c.c.$, propagating through a medium with a Kerr nonlinearity. The equation describing the evolution of the field as a function of y can be derived using the same procedure as that adopted for the case of the second-order nonlinearity. The result is:

$$\frac{dE}{dy} = i\frac{\omega}{c}3\chi^{(3)}|E|^2E. \tag{7.42}$$

Putting $E(y) = A(y)\exp[i\phi(y)]$, where $A(y)$ and $\phi(y)$ are real quantities, (7.42) splits into the two equations:

$$\frac{d\phi}{dy} = \frac{\omega}{c}3\chi_{Re}^{(3)}A^2, \tag{7.43}$$

and

$$\frac{dA}{dy} = -\frac{\omega}{c}3\chi_{Im}^{(3)}A^3, \tag{7.44}$$

describing the evolution of the nonlinear phase and of the field amplitude, respectively. Recalling that the light intensity I is proportional to A^2, and multiplying both sides of (7.44) by $2A$, it is found that $I(y)$ satisfies the equation:

$$\frac{dI}{dy} = -\beta_a I^2, \tag{7.45}$$

where

$$\beta_a = \frac{3\pi\chi_{Im}^{(3)}}{\varepsilon_o n^2 c\lambda}, \tag{7.46}$$

is known as the two-photon absorption coefficient. Equation (7.45) describes a decrease of intensity during propagation in an otherwise transparent medium. The fact that the derivative of the intensity is proportional to the intensity square suggests that the loss is due to the simultaneous absorption of two photons. If this is the case, β_a can be different from 0 only when the photon energy is larger than half the energy gap between ground state and excited state.

Integrating (7.45), it is found:

$$I(y) = \frac{I_o}{1 + \beta_a I_o y}, \quad (7.47)$$

Many experiments performed with semiconductors have indeed verified the behavior predicted by (7.47). As an example, the values of n_2 and β_a for two well-known semiconductors are reported in Table 7.2. Note that the energy of the incident photon is larger than half the bandgap energy of the two semiconductors, whereas, in the case of silica glass, the photon energy of 1.17 eV is too small to give rise to two-photon absorption.

Self-focusing. If the propagating beam has a finite cross-section, like a Gaussian spherical wave, then a new effect occurs, because the optical Kerr effect produces a transverse refractive-index profile that follows the intensity profile of the beam. If n_2 is positive, a greater index of refraction is induced on-axis than in the wings of the beam, creating a positive lens that tends to focus the beam. This effect is known as "self-focusing". The focal length of the self-induced Kerr lens can be calculated as follows. Assuming a radial intensity profile of the type $I_o \exp[-2(r^2/w_o^2)]$, the radially varying refractive index is given by:

$$n(r) = n_o + n_2 I_o \exp[-2(r^2/w_o^2)] \approx n_o + n_2 I_o\left(1 - 2\frac{r^2}{w_o^2}\right), \quad (7.48)$$

where this approximate expression is obtained by expanding the exponential in a power series, and then truncating the expansion after the first order.

If the thickness of the medium is L, then the radially dependent phase delay is:

$$\phi(r) = \frac{2\pi L}{\lambda}\left[n_o + n_2 I_o\left(1 - 2\frac{r^2}{w_o^2}\right)\right], \quad (7.49)$$

In order to understand the effect of a transverse distribution of the phase delay, the simplest strategy is to compare (7.49) with the transverse distribution created by a real lens having focal length f. The radial dependence of the phase delay due to a lens, as derived from (2.79) in Sect. 2.6, is:

$$\phi(r) = \phi_o - \left[\frac{2\pi}{\lambda}\frac{r^2}{2f}\right], \quad (7.50)$$

By comparing (7.49) and (7.50), the focal length of the Kerr lens, f_K, is found to be:

$$f_K = \frac{w_o^2}{4 I_o n_2 L}, \quad (7.51)$$

This lensing effect has a practical application: it is used to induce self-mode-locking in solid-state lasers, such as titanium sapphire lasers. The mechanism is the following. The multimode laser can operate either in a free-running regime, in which the modes have random phases, or in a mode-locking regime. If the number of modes is large, the peak intensity of the mode-locked pulses is much larger than the CW intensity of the free-running laser. It is assumed that the third-order nonlinearity of the active crystal gives rise to a Kerr lens in the mode-locking regime, but no Kerr lens in free running. The effect of the Kerr lens is that of focusing the laser beam at some point inside the cavity. If a pinhole is placed at that point, the free-running regime will experience large losses, whereas the mode-locking regime will be unaffected. As a consequence, the laser is forced to operate as a mode-locked oscillator. More generally, Kerr-lens mode-locking can be explained by keeping in mind the definition of cavity stability given in Sect. 3.8. If the cavity is designed to be stable in presence of the Kerr lens and unstable in absence of the Kerr lens, the laser will choose to operate in the mode-locking regime in presence of self-focusing. The technique is known as "Kerr-lens mode-locking".

When the material thickness becomes comparable to, or even larger than, f_K, then the description of the beam propagation becomes complicated because a nonlinear wave equation needs to be solved. By increasing the input power, a situation can be found in which self-focusing exactly balances diffraction, and so the beam propagates through the nonlinear medium without changing its profile. This phenomenon is known as "self-trapping". If the input power is further increased, the beam will catastrophically focus, causing optical damage inside the medium in many cases.

Self-phase-modulation. If an optical pulse of intensity $I(t)$ is propagating into a third-order nonlinear medium, the optical Kerr effect produces a time-dependent change in the refractive index experienced by the same beam. This phenomenon is known as "self-phase-modulation" (SPM). Neglecting for the moment any optical dispersion, the phase delay over a propagation length L is given by:

$$\phi(t) = \omega_o t - \frac{2\pi L}{\lambda_o}[n + n_2 I(t)] \tag{7.52}$$

SPM produces a broadening of the optical spectrum. The origin of broadening can be explained by noting that the instantaneous frequency ω_{inst}, defined as

$$\omega_{inst} = \frac{d\phi(t)}{dt} = \omega_o - \frac{2\pi L}{\lambda_o} n_2 \frac{dI}{dt}, \tag{7.53}$$

sweeps from a value smaller than ω_o at the pulse front, where dI/dt is positive, to a value larger than ω_o at the pulse tail.

In optical-communication systems, where short pulses propagate over long distances inside single-mode optical fibers, it is important to account for the effect of SPM on the pulse shape. A nonlinear equation describing the pulse evolution along the fiber can be derived from the wave equation by assuming that the pulse amplitude $A(y, t)$ is a slowly varying function of y and t ($dA/dy \ll \beta_1 A$ and $dA/dt \ll \omega_o A$), and truncating to the second order the power series expansion of the propagation constant β, as expressed in (6.19). The nonlinear equation obtained is then:

$$\frac{\partial A}{\partial y} + \beta_1 \frac{\partial A}{\partial t} + \frac{1}{2} i \beta_2 \frac{\partial^2 A}{\partial t^2} = i\gamma |A|^2 A, \tag{7.54}$$

where γ is a constant proportional to n_2:

$$\gamma = \frac{n_2 \omega_o}{c A_{eff}}, \tag{7.55}$$

and A_{eff} is the mode area. A typical value in the case of a single-mode silica fiber is: $\gamma = 2\ \mathrm{W}^{-1}\mathrm{km}^{-1}$.

In general the combined effect of dispersion and nonlinearity is for a broadening of the pulse both in time and frequency. However, if β_2 and n_2 have opposite signs, as is the case for a nonlinear medium exhibiting positive n_2 and anomalous group-velocity dispersion, it is possible to find a situation in which the two effects compensate each other. A pulse that can propagate unchanged is known as a temporal optical soliton, and has the form:

$$A_s(y, t) = A_{so} \mathrm{sech}\left(\frac{t - \beta_1 y}{\tau_s}\right) \exp^{i\kappa y}. \tag{7.56}$$

Equation (7.56) describes a pulse traveling at the group velocity $\beta_1{}^{-1}$, with a peak amplitude A_{so} and a duration τ_s that are mutually connected by the relation

$$A_{so}\tau_s = \sqrt{\frac{|\beta_2|}{\gamma}}, \tag{7.57}$$

and κ is given by:

$$\kappa = \frac{|\beta_2|}{2\tau_s{}^2}. \tag{7.58}$$

Note that the pulse duration τ_s can take any value, but the shorter the pulse the higher the peak intensity should be, according to (7.57). The pulse energy required, which is proportional to $A_{so}^2 \tau_s$, is inversely proportional to the pulse duration.

In principle, the pulse described by (7.56) could propagate without distortion for an arbitrary propagation length, and so could be of great interest for long-distance optical communications. However, the presence of effects that have been neglected in the derivation of (7.54), such as fiber losses and third-order dispersion, strongly diminishes the practical applications of solitons.

7.5 Stimulated Raman Scattering

When a light beam at frequency ν_p travels across a transparent medium, a usually small fraction of its power is scattered in all directions because of the presence of thermal fluctuations inside the medium. In addition to the main Rayleigh scattering contribution at ν_p, the spectrum of scattered light may also contain a much weaker contribution at the down-shifted frequency $\nu_{st} = \nu_p - \nu_v$, where ν_v is the frequency of a vibrational mode of the medium. This secondary process is known as "Raman scattering". The energy difference between the incident photon and the Raman scattered photon is used to excite a molecular vibration. The phenomenon occurs in molecular gases and liquids, and also in amorphous or crystalline solids. In the case of gases, Raman scattering can occur also in connection with molecular rotational levels.

Besides scattering events in which vibrational energy is transferred to the molecule, usually called Stokes processes, there are also events in which vibrational energy is taken from the molecule. This may happen if the molecule is already in an excited vibrational state. For these anti-Stokes processes, the Raman scattering event de-excites the molecule, and so the scattered photon has a frequency $\nu_{ast} = \nu_p + \nu_v$ (see Fig. 7.8). The number of anti-Stokes scattering events is smaller than that of Stokes scattering events because the number of excited molecules is smaller than that of non-excited ones. At thermal equilibrium, the ratio between the anti-Stokes and the Stokes intensity is given by:

$$\frac{I_{ast}}{I_{st}} = \left(\frac{\nu_{ast}}{\nu_{st}}\right)^4 \exp\left(-\frac{h\nu_v}{k_B T}\right). \tag{7.59}$$

Since, typically, $h\nu_v$ is of the order of 0.1 eV, whereas $k_B T = 0.025$ eV at room temperature, spontaneous Raman scattering is dominated by Stokes emission.

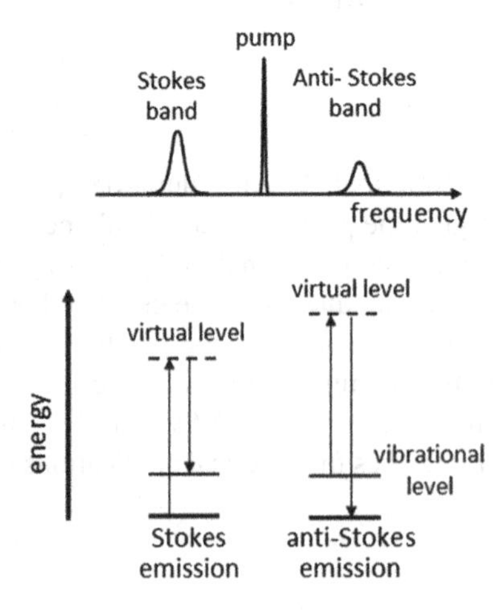

Fig. 7.8 Raman emission

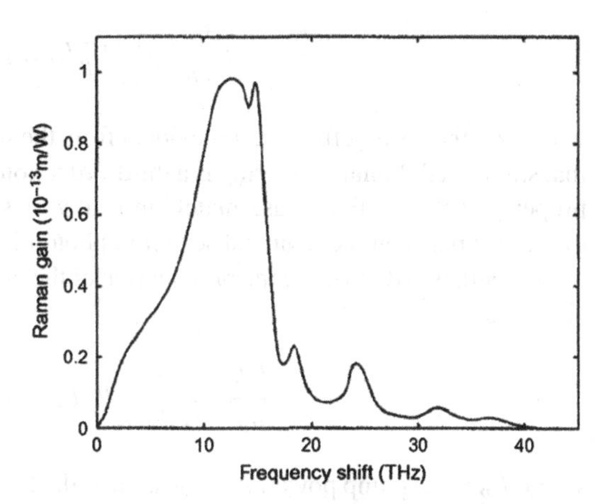

Fig. 7.9 Raman shift in silica glass

In the case of spontaneous Raman scattering, the Stokes photon is emitted with a random direction and a random phase. If the medium is simultaneously illuminated by a pump beam at frequency ν_p and a second beam at the Stokes frequency $\nu_{st} = \nu_p - \nu_v$, then there is a finite probability that a stimulated Raman scattering (SRS) process occurs, where the scattered Stokes photon is emitted with the same **k**-vector and the same phase of the incident Stokes photon. The stimulated process absorbs one photon from the pump, adds one photon to the Stokes field and, at the same time, excites a molecular vibration. Therefore, stimulated Raman scattering amplifies the field at ν_{st} at the expense of the pump power. Note that such an amplification method can be applied to beams of any frequency. In fact, provided that the medium is transparent to both pump and Stokes frequencies, it is not required that ν_p and ν_{st} be coincident with internal resonances of the atomic system, as is required for a standard light amplifier. To obtain Raman gain the pump field and the Stokes field need not necessarily be collinear, but the maximum effect is obtained in the collinear configuration, considering that the experiments are performed with Gaussian beams of finite transversal size. In the following, the discussion is limited to SRS processes occurring in single-mode optical fibers, where propagation of the two fields is necessarily collinear.

Fiber Raman amplifier. In the case of silica glass, Raman scattering offers a broad spectrum (see Fig. 7.9) centered around a shift of 13 THz, instead of a sharp line. This shift corresponds to the vibrational frequency of the $Si{-}O$ bond. The Raman peak is broadened because glass is a disordered medium, and so the vibrational frequency of the $Si{-}O$ bond is somewhat dependent on the random local environment. Assuming that a strong pump wave at frequency ν_p and a weak Stokes wave at a frequency $\nu_{st} \approx \nu_p - 13$ THz are launched into the fiber, the Stokes wave gets amplified through SRS. Since both pump wave and Stokes wave will be fiber modes, their transversal distribution is fixed during propagation (no diffraction!). Consequently, only the dependence of the field amplitudes on the propagation distance y has to be considered. The equation describing the evolution of the Stokes field is:

$$\frac{dE_{st}}{dy} = i\frac{3\omega_{st}}{cn_s}\chi^{(3)}E_p^* E_p E_{st} - \frac{\alpha}{2}E_{st}, \tag{7.60}$$

where α is the loss coefficient. It is evident from the structure of the gain term in (7.60) that stimulated Raman scattering is a third-order nonlinear process. A very important property of SRS is that phase-matching is always satisfied because the momentum difference between incident and scattered photon is taken up by the medium.

The equation describing the propagation of the optical power, P_{st}, as derived from (7.60), is the following:

$$\frac{dP_{st}}{dy} = \frac{g_R}{A_{eff}}P_p P_{st} - \alpha P_{st}, \tag{7.61}$$

where P_p is the pump power, A_{eff} the area of the fiber mode, and g_R the Raman-gain coefficient, defined by:

$$g_R = \frac{12\omega_{st}n_p}{\varepsilon_o c^2 n_{st}}\chi_{Im}^{(3)}. \tag{7.62}$$

Note that the Raman gain depends on the imaginary part of $\chi^{(3)}$. This is not surprising because processes in which there is an exchange of energy between wave and medium generally depend on the imaginary part of the electric susceptibility, as has already been seen in the case of two-photon absorption.

Assuming no pump-depletion, i.e. a constant P_p, the integration of (7.61) gives the Raman gain in a fiber of length L:

$$G = \exp\left[\left(\frac{g_R}{A_{eff}}P_p - \alpha\right)L\right] \tag{7.63}$$

For instance, at $\lambda = 1.55\ \mu$m, taking $g_R/A_{eff} = 3.2 \times 10^{-3}\ \mathrm{m^{-1}W^{-1}}$, $P_p = 2$ W, $\alpha_o = 0.2$ dB/km, which means $\alpha = 0.46 \times 10^{-4}\ \mathrm{m^{-1}}$, $L = 500$ m, it is found a Raman gain of about 25. Signal amplification by SRS is commonly used in optical communication links.

So far only stationary beams have been considered. What happens with short pulses? It is found that the Raman gain has a build-up time that is related to the spontaneous decay time of molecular vibrations. The precise value depends on the specific vibrational transition, but a typical value is $\approx$1 ps. Therefore, for pulses shorter than about 1 ps, the Raman gain becomes smaller and is essentially negligible for femtosecond pulses.

If only the pump beam is present at the fiber input, then spontaneous Raman scattering provides a weak signal that can act as a seed for growth of the Stokes wave. If the first Stokes wave becomes sufficiently powerful, then it can in turn generate the second Stokes, and so on. As a consequence, a large fraction of the input power can be transferred to the Raman lines. Furthermore, it should be noted that stimulated Raman scattering may be either forward- or backward-propagating in the fiber.

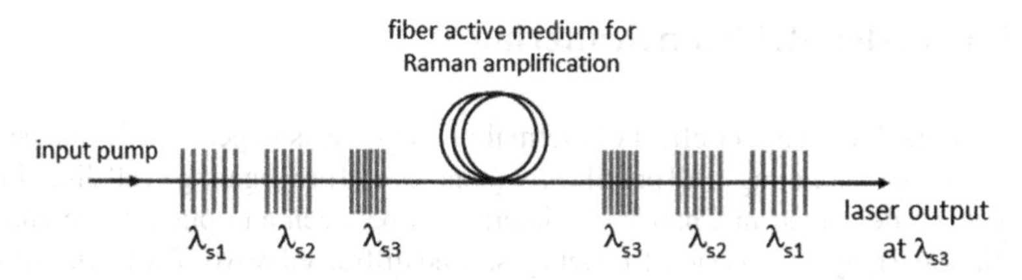

Fig. 7.10 Scheme of the Raman fiber laser based on a nested cavity. In a laser designed to emit the third Stokes wave, all the FBGs have high reflectivity, except for the λ_{s3} output coupler

Backward-propagating stimulated scattering process can result in a severe attenuation of the forward propagating wave. The spontaneous generation of a stimulated Stokes wave is detrimental for Raman fiber amplifiers, and so it is important to know what is the critical pump power, P_{cr}, above which significant stimulated effects are generated even in absence of an input signal. It can be shown that

$$P_{cr} \approx 16\frac{A_{eff}\alpha}{g_R}. \tag{7.64}$$

A typical value for a single-mode fiber is: $P_{cr} \approx 1.5\,\mathrm{W}$.

Fiber Raman laser. SRS is a very interesting method not only to amplify signals, but also to generate new laser wavelengths by converting the Raman fiber amplifier into an oscillator. The availability of high-power diode lasers as pump sources and the development of high reflectivity fiber Bragg gratings makes it possible the realization of nested cavities in which several Stokes lines can simultaneously resonate. In this configuration the pump power can be very efficiently converted to the highest-order Stokes line.

A Fabry-Perot nested cavity is shown in Fig. 7.10. The feedback elements are fiber Bragg gratings (FBGs), written directly on the core of the fiber, and placed at the input and output sides of the fiber. This nested scheme can generate a cascade of n frequencies through stimulated Raman scattering: each intermediate Stokes line at wavelength λ_{sk} oscillates in a cavity made by a pair of high-reflectivity FBGs, and can be efficiently converted into the λ_{sk+1} Stokes line. The cascade is terminated at λ_{sn} by a low-reflectivity FBG, which couples a significant fraction of the λ_{sn} Stokes power to the output. An additional grating centered at the pump wavelength ω_p yields a double-pass pumping scheme. The stimulated Raman effect couples waves which are co- or counter-propagating in the same way. As an example, the pump source could be a CW diode-pumped cladding-pumped Yb^{3+} fiber laser at 1117 nm. By using such a pump, a cascade of five (1175, 1240, 1315, 1395 and 1480 nm) orders of Stokes can produce the 1480 nm wavelength. The efficiency of such a laser can be very large: typically 50 % of the input power at 1117 nm can be converted into output power at 1480 nm. Such a source can be very useful as a pump for high-power and remotely-pumped erbium amplifiers.

7.6 Stimulated Brillouin Scattering

In a dense medium another effect of thermal fluctuations is to generate spontaneous acoustic waves oscillating at all possible frequencies and propagating in all directions. As has been described in Chap. 4, the interaction between an optical wave and an acoustic wave may give rise to a frequency-shifted diffracted wave. The phenomenon by which an incident beam is diffracted by a randomly excited acoustic wave inside condensed matter is known as Brillouin scattering.

Considering a plane wave with frequency ν_i and wave vector $\mathbf{k_i}$ incident on the medium, Brillouin scattering is observed in all possible direction. Selecting light scattered at an angle α with respect to the direction of $\mathbf{k_i}$, the observed spectrum of the scattered light is found to contain three peaks, one at ν_i and two at the frequencies $\nu_i \pm \nu_s$, where ν_s is the frequency of the spontaneous acoustic wave that produces a diffracted beam in the chosen direction. Taking $\mathbf{k_d}$ as the wave vector of the photon diffracted at angle α, then the wave vector $\mathbf{k_s}$ of the acoustic wave involved in the process is given by (4.43) (see also Fig. 4.12). The peaks at the frequencies $\nu_i \pm \nu_s$ are both found in the spectrum because both orientations of $\mathbf{k_s}$ are present inside the medium. The modulus of $\mathbf{k_s}$ is derived from (4.40), taking into account that the angle θ of (4.40) coincides with $\alpha/2$. The corresponding acoustic frequency is:

$$\nu_s = \frac{k_s}{2\pi u_s} = \frac{k_i}{\pi u_s}\sin(\alpha/2) \tag{7.65}$$

Similarly to the case of SRS, it is possible to have stimulated Brillouin scattering (SBS) at the frequency $\nu_i - \nu_s$. The SBS wave is usually observed in the backward direction ($\alpha = \pi$), because the maximum overlap between the incident and the scattered wave occurs for that direction. In the forward direction $k_s = 0$, so there can be no Brillouin shift.

While the material nonlinearity of silica is actually not very high, the small effective mode area and long propagation length used in fibers tend to strongly favor nonlinear effects. For silica fibers, the Brillouin frequency shift is of the order of 10 – 20 GHz. The dependence of the Brillouin frequency shift on the material composition and, to some extent, on the temperature and pressure of the medium can be exploited for fiber-optic sensors.

If only the pump beam is present, then spontaneous Brillouin scattering provides a weak signal at $\nu_i - \nu_s$ that can act as a seed for growth of a stimulated Brillouin scattering wave. Inside optical fibers the spontaneous generation of SBS is frequently encountered when narrow-band optical signals (e.g. from a single-frequency laser) are amplified in a fiber amplifier, or have just propagated through a passive fiber. This effect is usually undesirable, and so the optical setup is designed to avoid the generation and propagation of SBS.

Problems

7.1 The frequency of a Nd-YAG laser output ($\lambda = 1064$ nm) is to be doubled in a lithium niobate crystal by type-I birefringence phase matching. Knowing that, for $LiNbO_3$, $n_o(\lambda = 1064\ \text{nm}) = 2.2347$, $n_e(\lambda = 532\ \text{nm}) = 2.2316$, and $n_o(\lambda = 532\ \text{nm}) = 2.3301$, calculate the phase-matching angle between the optical axis and the direction of the incoming beam.

7.2 Calculate the second-harmonic conversion efficiency for type-I harmonic generation in a perfectly phase-matched 2-cm long KTP crystal with an incident beam at $\lambda = 1064$ nm having an intensity of 3 MW/cm^2. (For KTP, $n_\omega = n_{2\omega} \approx 1.8$, and $\chi^{(2)} = 13\,\text{pm/V}$).

7.3 The generation of the second-harmonic of 1064-nm light is performed by using a periodically poled lithium niobate crystal ($n_e = 2.1550$ at $\lambda = 1064$ nm and $n_e = 2.2316$ at $\lambda = 532$ nm). Calculate the period of periodical poling required for phase matching of second-harmonic generation.

7.4 A parametric amplifier uses a 2-cm long KTP crystal ($n_1 \approx n_2 \approx 1.77$, $\chi^{(2)} = 10$ pm/V) to amplify light of wavelength 580 nm in a collinear wave configuration. The pump wavelength is $\lambda = 400$ nm and its intensity is 4 MW/cm^2. Determine the gain coefficient and the overall gain.

7.5 A Q-switched Nd-YAG laser produces an output pulse containing 10 mJ of energy with a pulse duration of 10 ns. The pulse generates an optical Kerr effect in a 2-cm thick carbon disulfide ($n_2 = 3.2 \times 10^{-14}\ \text{cm}^2/\text{W}$) cell. Assuming that $w_o = 100\ \mu\text{m}$, calculate the focal length due to the self-focusing effect.

7.6 A 1540-nm 2-ns laser pulse impinges on a 1-mm thick GaAs plate. Assuming a spot size of 0.2 mm and a two-photon absorption coefficient $\beta_a = 10.2$ cm/GW, calculate the transmittance of the plate for the two cases in which the pulse energy is 1 mJ and 100 μJ.

7.7 A laser pulse at $\lambda_o = 1064$ nm propagating into an optical fiber excites a cascade of stimulated Raman scattering processes. Assuming that the Raman frequency shift is 13 THz, calculate the wavelength of the second Raman Stokes line, λ_{s2}.

Chapter 8
Applications

Abstract This chapter concisely describes the most important applications of photonics, which are based on the optical methods and devices discussed in Chaps. 1–7. Section 8.1 deals with the applications to information and communication technology. After introducing the design principles of digital information transmission and recording, the structure and properties of an optical communication link are presented. This is followed by a description of optical memories and of the prospects of integrated photonics. Section 8.2 is dedicated to optical metrology and sensing, including the optical radar, the laser gyroscope, and fiber-optic sensing of temperature, strain, and electric current. The achievements and the potentialities of laser materials processing are reviewed in Sect. 8.3, with some examples of the many possible biomedical applications treated in Sect. 8.4. The last three sections concern topics of very wide interest, with an impact on everyday life: displays, lighting and photovoltaic energy conversion. In all of these, photonic methods and devices play an increasingly important role, even if lasers are not directly involved.

8.1 Information and Communication Technology

Photonics pervades all the aspects of the information and communication technology. Optical-fiber transmission of data, voice, and video is the basic technology for long-haul communications and internet traffic, and is now also reaching individual users directly via fiber-to-the-home broadband networks. Optical discs are used for distributing digitally encoded music and films, and for storing large amounts of data. Printing and photocopying machines rely on lasers. High-speed data processing will benefit from the development of integrated photonic circuits in the near future.

Before describing specific applications, it is useful to discuss some general aspects concerning digital information recording and transmission.

A written text is a combination of alphabetical and mathematical symbols. Each symbol is identified by a sequence of 7 binary units, known as bits, allowing $2^7 = 128$ combinations (ASCII code). As an example, a page of 2000 characters corresponds to a sequence of 14000 bits, that is, 14 kb.

V. Degiorgio and I. Cristiani, *Photonics*, Undergraduate Lecture Notes in Physics,
DOI 10.1007/978-3-319-20627-1_8

An acoustic signal (voice and sound) induces a vibration of a receiving membrane that is converted into an electric signal by a transducer. The electric signal is digitalized by an analog-to-digital converter, which transforms an analog function of continuous time into a numeric sequence of discrete times. The minimum sampling interval is determined according to the sampling theorem of Nyquist and Shannon, which states: if a function of time $f(t)$ has a spectral width $\Delta\nu$, it is completely determined by giving its values at a sampling rate of $2\Delta\nu$. If the digitalization makes use of $2^8 = 256$ levels, then each sample is given by a sequence of 8 bits. The typical bit rate required for a digital audio signal is about 60 kb/s.

A video signal consists in a sequence of 20 images per second. An image is a matrix of about $700 \times 500 = 350{,}000$ pixel. This number needs to be multiplied by a factor of 3 to account for the RGB (red, green, blue) color code. The required bit rate can be considerably reduced by using very effective compression techniques.

8.1.1 Optical Communications

An optical communication system uses a transmitter, which encodes a message into an optical signal, a waveguide, which carries the signal to its destination, and a receiver, which converts the optical bit stream into the original electrical form. Optical fiber is the standard type of waveguide for optical communications. Free-space optics is useful for communications between spacecrafts, but is not of much interest for terrestrial links, because its range is limited and the quality of the link is too highly dependent on atmospheric factors. The first fiber-optic communication systems have been installed in the 1980s. In particular, the first transatlantic submarine fiber-optic cable was laid down in 1988 in order to connect the United States and Europe. Two years later a submarine connection linked the United States to Japan. The world-wide deployment of fiber systems took place in the 1990s, following the development of single-mode semiconductor lasers and optical fiber amplifiers. The carrier wavelength, which must be close to the minimum-attenuation wavelength of silica-fibers, varies from 1520 to 1620 nm (185 to 200 THz). The interval 1530–1570 nm is known as the C band, while the L band corresponds to 1570–1620 nm.

A scheme of a point-to-point optical-communication system is shown in Fig. 8.1. The most common light source for high-bit-rate optical communications is the single-mode DFB semiconductor laser. The optical signal to be transmitted is a stream of bits, created by an external modulator or by direct modulation of the laser pump current. The signal encoding is typically simple amplitude modulation, in which the presence of a pulse means 1 and the absence means 0. The direct-modulation method is simple, but the generated pulses are usually not transform-limited, and so are more strongly degraded by optical dispersion in long distance communications. The external modulation by an electro-optic modulator of the type described in Chap. 4 not only ensures a better pulse quality, but also offers a greater flexibility in the choice of the modulation format. An interesting encoding method, which may substitute amplitude modulation, is the phase modulation method known as "differential phase

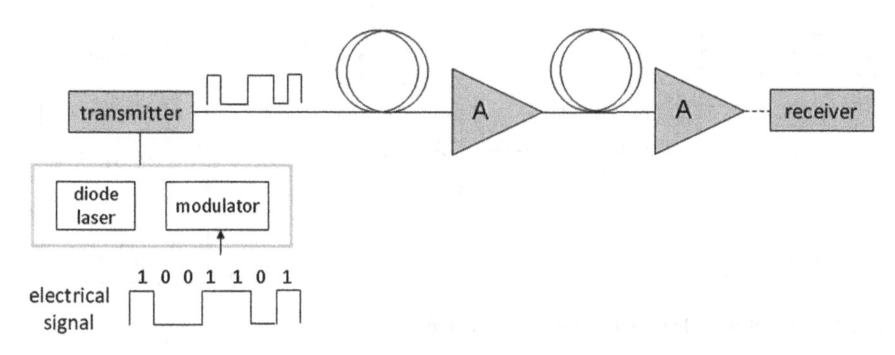

Fig. 8.1 Scheme of a point-to-point fiber-optic link. *A* optical amplifier

shift keying" (DPSK). Phase modulation formats allow for more than two phase levels per single bit, and thus provide an increase in bit-rate, obtainable without changing the physical structure of the link.

The existing links are using single-mode fibers of type SMR or DS. The effects of optical dispersion can be compensated for by inserting DIS-CO fibers into the link. These have a large dispersion parameter, D, with a sign opposite to that of SMR or DS fibers. Another possibility for compensation is through the use of chirped fiber Bragg gratings, which are specifically designed to provide a large dispersion with the appropriate sign.

Since the optical signal is attenuated during propagation along the fiber, there is a need for periodic signal regeneration. The amplification is usually provided by the erbium-doped fiber amplifiers (EDFAs), described in Sect. 6.6. Semiconductor optical amplifiers (SOAs) can also be used for signal regeneration. These enjoy different advantages and disadvantages with respect to EDFAs. The main advantages are electrical pumping, small size, low cost, and a central wavelength that is selectable through the choice of material. The main disadvantages are the low saturated output power (which means limited linearity), the sensitivity to signal polarization and to temperature, the high insertion loss, and high noise. In particular, since the gain and refractive index of a SOA can evolve on the picosecond time scale, the amplified pulse may be distorted in both amplitude and phase. In contrast, the gain dynamics of EDFAs is slow, and so picosecond pulses are linearly amplified with a gain determined only by the average pump power. As discussed in Sect. 7.6, signal regeneration can be also performed using Raman fiber amplifiers, which are based on stimulated Raman scattering. Raman fiber amplifiers can cover a very broad bandwidth. They are often used in optical links, as amplifiers complementary to EDFAs.

At the receiver a photodetector, such as a *p-i-n* photodiode or an avalanche photodiode, is used for optical-to-electrical conversion. The demodulation re-creates the electrical bit stream used at the transmitter to modulate the optical carrier. In every communication link there is some unavoidable noise added during transmission and at the receiver, and this leads to errors. The performance of a digital communication system is measured by the probability of error per bit, known as the "bit error rate"

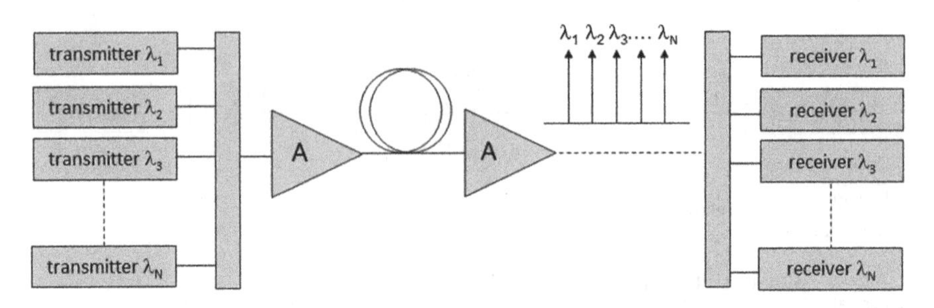

Fig. 8.2 Wavelength division multiplexing technique

(BER). The BER is required to be $<10^{-9}$, which means one wrong bit every billion bits. All receivers need a certain minimum power to operate reliably. This power level is known as the receiver sensitivity.

One way of increasing the transmission bandwidth of an optical link is to employ the technique of "wavelength division multiplexing" (WDM). In this scheme, signals with different carrier wavelengths are mixed together (i.e. multiplexed) and then transmitted simultaneously down a fiber, according to the scheme shown in Fig. 8.2. This technique increases the capacity of an optical fiber system without changing the fiber and without increasing the bit rate in the single carrier. In order to minimize cross-talk, the frequency separation between two adjacent carriers (called "channels" in the optical-communication jargon) must be greater than the modulation bandwidth used by a single channel. A typical situation is that of a 40-channel WDM system, using 40 lasers, all operating in the C band, with a wavelength spacing of 0.8 nm (corresponding to a frequency spacing of 100 GHz). If the bit rate of each channel is kept around 10 Gb/s, then interference among different channels is negligible. This system has the same capacity of a single channel that transmits at 400 Gb/s. At present, WDM systems with a capacity of several Tb/s are available. WDM can play an important role in a network where a number of users are connected to a central hub. Each user is assigned a particular transmission wavelength. The individual wavelength required by each user is isolated by using a narrow-band optical filter.

8.1.2 Optical Memories

In an optical memory, data are stored in a digital format on an optical disc, and can be read-out by a laser beam. An optical disc is a flat circular disc on which binary data (bits) are encoded in the form of pits (binary value of 0, due to lack of reflection when read) and lands (binary value of 1, due to a reflection when read) on a special material on one of its flat surfaces. The diameter of a single pixel and the pixel separation are about 1 μm. A laser source writes one bit of data at a time on the disc, typically through a thermal mechanism. The pixels are written along a spiral developing from an inner radius to an outer radius.

A variety of different materials and recording mechanisms can be used. Both write-once read-many (WORM) and magneto-optic read/write discs are presently available. In the case of WORM discs, a focused laser beam heats a spot on the surface layer during the recording stage. The local reflectance of the surface is irreversibly changed, either by creating a pit or by modifying the surface properties. In the case of the read/write disc, light absorption reversibly transforms the crystalline structure of the surface layer (typically made by the *AgInSbTe* or *GeSbTe* compounds) to produce a large difference in reflectivity through a phase transition. If the material is magneto-optic, such as some rare-earth transition-metal alloys of *TbFeCo*, the effect of the phase transition is to rotate the polarization of reflected light. The rotation is detected by crossed polarizers.

The first generation of the compact discs (CDs), developed in the mid-1980s, were designed to hold up to 75 min of music which required 650 Mb of storage.

Optical discs are usually between 7.6 and 30 cm in diameter, with 12 cm being the most common size. A typical disc is about 1.2 mm thick, while the track pitch (distance from the center of one track to the center of the next) is typically 1.6 μm.

Optical discs are most commonly used for storing music (e.g. for use in a CD player), video (e.g. for use in a Blu-ray player), or data and programs for personal computers. For computer data backup and physical data transfer, optical discs are gradually being replaced with faster, smaller solid-state devices, especially the USB flash drive.

The objectives used to focus the light beams are characterized by their numerical aperture (NA), defined in Sect. 2.2.9. When a light beam is focused by an objective with a given NA, the diameter of the obtained focal spot is:

$$d_f = \frac{a\lambda}{\mathrm{NA}}, \tag{8.1}$$

where a is a numerical constant of the order of 1. The area used to record one bit of information is $2{d_f}^2$. If $\lambda = 0.8\ \mu\mathrm{m}$, it is possible to memorize 150 Mb over 1 cm^2. The basic digital memory unit is the byte (B), which is equivalent to 8 bit. To give an example of the used notation: 10^6 Byte = 1 MegaByte = 1 MB.

An important technical aspect concerning the design of optical disc readers is that the distance between lens and disc surface must remain constant to within an accuracy of the order of the Rayleigh length (a few microns). This requires a sophisticated real-time active stabilization system.

The most common optical discs have the following properties.

Compact Disc. A Compact Disc (CD) uses a *GaAlAs* semiconductor laser, emitting infrared light of 780 nm, for writing and reading. The laser beam is focused with an objective that has a NA = 0.45–0.50. The size of the focal spot on the disc surface is 1.6 μm. The CD capacity is of 680 MB.

Digital Versatile Disc. A Digital Versatile Disc (DVD) uses a *GaAlAs* semiconductor laser, emitting red light of 650 nm, for writing and reading. The laser beam is focused with an objective that has a NA = 0.60–0.65. The size of the focal spot on the disc surface is 1.1 μm. The DVD capacity is of 4.7 GB, and can be doubled by using both disc surfaces.

Blu-Ray. A Blu-Ray disc uses a $GaInN$ semiconductor laser, emitting light of 405 nm (notwithstanding the name of the disc, this is violet light!), for writing and reading. The laser beam is focused with an objective that has a NA = 0.85. The size of the focal spot on the disc surface is 0.48 μm. The DVD capacity is of 25 GB, and can be doubled by using both disc surfaces.

The capacity of optical memories could be significantly increased by using "volume" storage instead of "surface" storage. Such a possibility is actively being investigated, but, at present, no commercial products are available.

8.1.3 Integrated Photonics

Optical systems, as they have been described throughout this book, contain various discrete elements, each made using different materials and technologies, with dimensions several orders of magnitude larger than the wavelength of light. A major challenge for the future will be the development of high-bandwidth, low energy consumption optical circuits at low cost. This will be possible provided optical systems evolve towards combining all devices and components for the generation, manipulation, and detection of light onto a single substrate, i.e. a single "chip". Such a chip is the optical analogue of an electronic integrated circuit, and should possibly be based on silicon, in order to exploit the technologies already available in microelectronics. Silicon is abundant, non-toxic and environmentally friendly. Since silicon is transparent to light of wavelength larger than 1.15 μm, silicon integrated photonics operates in the infrared. Photonic integration will play a key role in reducing cost, space and power consumption, and improving flexibility and reliability.

An important application of integrated photonics would be that of realizing on-chip processors for short-distance optical communications, i.e. the so-called "local area networks" (LAN). Another very important area concerns optical interconnections, through transceivers, which are devices that convert electric signals into optical signals and transport the optical signal from one device to another device. Optical data transportation is immune from induction effects, and may also reduce energy dissipation through Joule heating. Indeed, it is believed that a major obstacle for the realization of next-generation supercomputers will be energy dissipation due to data transfer among microprocessors and memories. To save energy, power-consuming parts of the electrical interconnects should be replaced by optical interconnects.

Research and development in the field of silicon photonics is being very actively pursued. Promising results have been obtained in the design and fabrication of passive components, such as waveguides, mirrors, and filters. Various types of silicon modulators have also been tested. The main difficulties concern light amplification, because electron-hole recombination is not radiative in bulk silicon, as discussed in Chap. 5.

8.2 Optical Metrology and Sensing

In many industrial processes, tests of structural stability, and environmental studies, there exists a need to monitor variables such as displacement, strain, temperature, pressure, flow rate and chemical composition. Sensing is also important because the data acquired in real-time may be used as input signals for automatic systems of industrial production and environmental control.

Optical sensing is useful in a variety of areas, because of its high accuracy, immunity to electromagnetic disturbances, little invasiveness, and the possibility to operate in a hostile environment. In order to illustrate the potential of optical sensing, a few applications are below described, without any pretension to be exhaustive.

8.2.1 Measurements of Distance

To measure distances, light is emitted towards and reflected (or scattered) from an object. Using the speed of light, and the measured time-of-flight (or phase delay) of the reflected light, it is possible to calculate the distance between laser emitter and target object. Different optical methods are available, depending on the range of distances and on the required accuracy.

Optical radar. Distance measurement with a pulsed-laser radar is based on the time-of-flight principle: the distance is determined from the time needed by a short light pulse to travel from the sender to the object, and back. In a range from a few meters to several hundred meters, this is the method of choice for determining distances. In the case of pulsed laser radars, every measurement unambiguously corresponds to the distance to a target object, even if multiple targets are hit by the laser beam. A laser radar can process fast variations in distance without dynamic errors. This is necessary for fast scanning of scenes with pronounced distance variations, or measurements of rapidly changing distances. Laser diodes are generally employed as the light sources. Fast and sensitive avalanche photodiodes (APD) provide the detectors.

The largest distance ever measured by an optical radar is the distance between the Earth and the Moon, L_{EM}, which is about 3.8×10^5 km, corresponding to a delay τ of 2.5 s. Of course, L_{EM} is not constant because the Moon's orbit around the Earth is elliptical. It is instructive to discuss the problems encountered in making such a measurement. The direct signal due to back-scattering from the lunar surface is too weak to be detected by the receiver, so it was necessary to place a high-reflectance mirror on the Moon. A normal mirror would be of little use, because it cannot reflect back in exactly the opposite direction, independently from the value of the incidence angle. The mirror brought by the astronauts of the first Apollo mission in July 1969 consisted of a two-dimensional matrix of corner cubes (see Sect. 2.2.6), with an area of $A_s \approx 1$ m^2. Assuming that the laser beam has a divergence of $\theta_o = 10^{-4}$ rad, and using (1.40) and (1.41), the radius of the illuminated area on the lunar surface is given by $w_L = L_{EM}\theta_o = 38$ km. The power fraction intercepted by the installed mirror is $A_s/(\pi {w_L}^2)$. If the back-reflected beam is collected over a receiver area equal to A_s, the total fraction of power received by the detector is:

$$\frac{A_s}{4\pi {w_L}^2} \approx 5 \times 10^{-11}. \tag{8.2}$$

Such a small value would make the measurement difficult. However, the power ratio of (8.2) can be increased by enlarging with a telescope the radius of the laser beam.

Time-of-flight sensing in pulsed mode is the dominant technique for distances longer than 50 m and is routinely used for many civilian applications of range sensing, such as mapping and surveying.

An alternative approach to pulse illumination is to use amplitude modulated continuous light. In this case a phase shift in the modulation signal is measured between the launched and the returned light, and the time-of-flight is determined by dividing the phase shift by the modulation frequency ω_m. The measurement may present an ambiguity because the measured phase shift is a periodic function of the distance. This ambiguity can be eliminated by appropriately choosing the modulation frequency or by measuring at more than one modulation frequency.

Interferometric method. The target is illuminated by a continuous laser beam, exhibiting a very good amplitude and phase stability. The back-reflected beam is superposed with a fraction of the incident beam using a Michelson interferometer, as shown in Fig. 8.3. The light intensity at the detector is proportional to the square modulus of the sum of the two fields. By using the same approach utilized to describe the Mach-Zehnder interferometer in Sect. 4.1.2, it can be shown that the output signal from the photodetector is proportional to $\cos^2(2\pi \Delta L/\lambda)$, where ΔL is the difference between the two arms of the interferometer. There is ambiguity in such a result, because the addition of an arbitrary number of half-wavelengths to

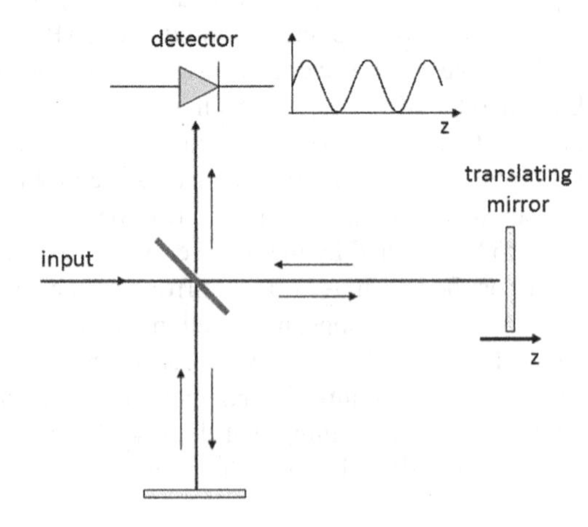

Fig. 8.3 Michelson interferometer

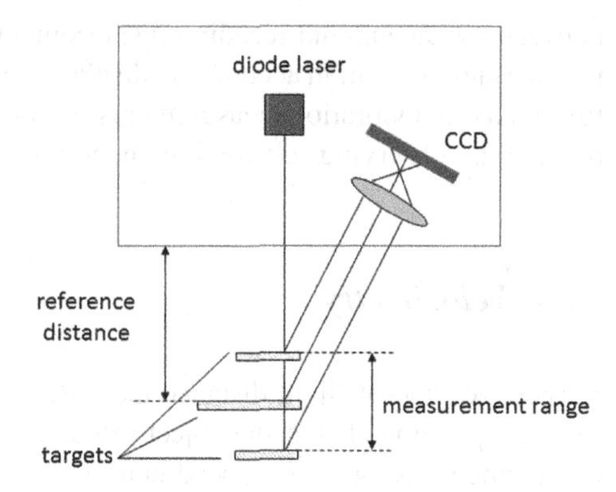

Fig. 8.4 Distance measurement through triangulation

ΔL would not change $\cos(\Delta\phi)$. Generally, interferometry is more appropriate for displacement monitoring than for absolute distance measurement, though multiwavelength and scanning interferometry permit absolute distance measurement as well. Interferometric techniques are particularly suitable to detect small displacements or small-amplitude vibrations in structural and mechanical engineering investigations or in geodetic surveys.

Triangulation. A collimated light source is used to illuminate the target whose distance is to be measured. A camera lens optics, laterally displaced from the light source, images the laser spot on the object onto a position-sensitive detector (typically a linear array of CCD detectors), as shown in Fig. 8.4. By knowing the distance between source and lens and measuring the position of the image of the light spot on the array of detectors, the distance of the object can be deduced. It has been reported that a similar method was used around 600 BC by the Greek mathematician Thales of Miletus to measure the height of the pyramids of Giza and also to determine the distance to a ship at sea. Triangulation is used to fix the focusing distance in photographic cameras. Triangulation sensors offer the advantages of a low price and fast measurement (data acquisition rates of tens or hundreds of kilohertz are possible). Red laser light is normally used, but the eye-safe 1.5 μm infrared light can also be employed. The useful range of distances for commercial instruments is from one centimeter to a few meters. Adding the capability to scan the position of the triangulation laser spot in the object plane opens up the possibility for object shape sensing.

Self-mixing interferometry. Self-mixing interferometry uses a single-mode laser, typically a simple semiconductor laser, and exploits the fact that any back-reflected signal from the target that reenters the laser cavity will modulate the laser output. This occurs because the target and the diode internal cavity can be regarded as forming a compound cavity. Because of feedback from the object reflection, the laser power output oscillates as the target object is displaced; one oscillation period corresponds to a target displacement of half the wavelength. If the feedback is moderately strong, then the oscillations are asymmetrically distorted, and it is possible to distinguish between advancing and receding displacement. The self-mixing technique is inexpensive and offers high accuracy in displacement measurement, and can be extended to velocity and vibration measurements. Absolute distance measurements have been reported by observing self-mixing under current modulated operation.

8.2.2 Velocimetry

Successive time-of-flight distance measurements can determine the rate of change in object position, that is, the object velocity. A typical application is in traffic speed enforcement, using a laser speed gun, which repeatedly measures the distance to a moving car at a high repetition rate.

Similarly, successive distance measurements can be used to monitor periodic motion such as vibrations. This is subject to the following restrictions: (i) The amplitude of motion is larger than the distance resolution of the measurement, and (ii) The period of motion is significantly longer than the time required for each individual distance measurement.

Laser Doppler velocimetry. An alternative optical technique for direct measurement of velocity and vibration is optical Doppler sensing. This deduces the velocity of moving objects by detecting the Doppler frequency shift of light reflected, or scattered, from a moving object. In optical Doppler sensing the light source is a CW laser, and detection is usually by a heterodyne technique. The detector simultaneously receives the launched frequency and the Doppler shifted frequency from the moving object, and the addition of these two waves produces a beat at the difference frequency (easily seen on an oscilloscope) at the detector output. Fourier transform analysis of the detector signal, performed by an electronic spectrum analyzer, gives the frequency shift.

The general expression of the Doppler shift is given by (4.41). In the particular case of back-reflection (or back-diffusion), $\mathbf{k_d} = -\mathbf{k_i}$, the Doppler shift is:

$$\Delta\nu_D = \nu_d - \nu_i = 2\frac{u}{c}\nu_i, \tag{8.3}$$

where ν_i is the laser frequency, ν_d the frequency of the back-reflected beam, and u the projection of the target velocity on the beam direction. The projection is taken as positive if it is directed toward the emitter. As an example, using a laser beam at 600 nm, corresponding to $\nu_i = 5 \times 10^{14}$ Hz, and considering an object that travels toward the emitter at a speed of 36 km/h $=$ 10 m/s, a positive Doppler shift of $\Delta\nu_D \approx 33$ MHz is calculated. It should be noted that the sign of the Doppler shift is lost when measuring the beat frequency at the detector output. A simple technique that allows for the recovery of the velocity direction consists of the introduction of a controlled frequency shift (by using one of the modulation methods described in Chap. 4) on that part of the laser beam which is sent to the detector. If the laser frequency is shifted by ν_s, then the measured shift becomes $\Delta\nu_D = \nu_d - \nu_i - \nu_s = 2(u/c)\nu_i - \nu_s$. In this case, a positive u gives a shift different from that of a negative u, and so the ambiguity is removed.

Laser Doppler velocimetry is a non-invasive method with useful applications in fluidodynamics, where measurements of the velocity field in the laminar or turbulent motion of a liquid inside a duct can be performed, and in aerodynamics, where wind tunnel flow measurements can provide unique insights into complex aerodynamic phenomena. In addition, laser velocimetry has health care applications, such as arterial blood flow measurements. In all these cases, the return signal does not come from the fluid itself, but is rather due to scattering from suspended particles following the local fluid motion. If suitable particles are not already present in the fluid, then seeding of controlled-size (typically of diameter around 1 μm) particles can be carried out.

The most common technique utilized in fluidodynamics is that of dividing the laser beam in two parts, which cross each other in the fluid region where the velocity has to be measured, as shown in Fig. 8.5. Calling θ the angle between the propagation directions of the two beams, a periodic set of parallel interference fringes is formed inside the intersection region, with a fringe spacing given by:

$$s = \frac{\lambda}{2\sin(\theta/2)}. \tag{8.4}$$

A particle suspended in the flowing fluid produces, when crossing the periodic set of fringes, a scattered light intensity that is proportional, at each point to the local light intensity. Therefore, the light intensity scattered by the flowing particle is oscillating in time, with a period equal to the flight time between two consecutive fringes (i.e. between an intensity maximum and the next intensity maximum). The oscillation period is $T_D = s/u$, where u is the velocity component in the direction perpendicular to the fringe planes. Since T_D is independent from the observation

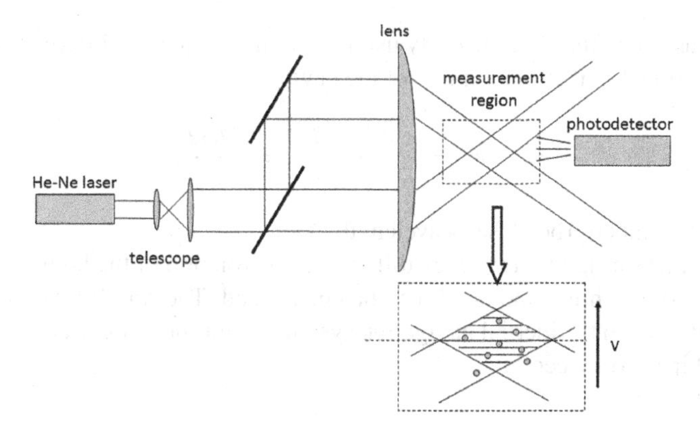

Fig. 8.5 Scheme of a laser Doppler velocimeter

direction, the scattered light signal can be collected with a large aperture lens. The period T_D is derived by analyzing the oscillations in the electric current output of the photodetector. If the full velocity vector is needed, then the measurement must involve three sets of fringes with different orientations.

Gyroscope. Rotation sensors, known as gyroscopes, are essential instruments for aircraft navigation. Optical gyroscopes rely on the Sagnac effect, which consists in a phase shift induced between two light beams traveling in opposite directions round a circular path when the platform on which the optical components are fixed is rotating about an axis perpendicular to the platform. In a single-mode ring laser cavity, the two oppositely directed traveling beams have the same frequency, which is inversely proportional to optical path length in a cavity round trip, as discussed in Sect. 3.8. Thus any physical mechanism that causes the optical paths to differ for the two beams, results in the oscillation frequency being different. To calculate the expected phase shift, assume that the cavity is a circular ring of radius R. Neglecting the correction due to the fact that part of the path may not be in air, if there is no rotation the optical path corresponding to one turn is equal to $2\pi R$ for both beams. Therefore both beams will return to the starting point in a time $T = 2\pi R/c$. If, however, the ring is rotating in a clockwise direction at a rate Ω rad/s, the clockwise beam will arrive at the starting point in a time T^+ longer than T because the optical path is now $2\pi R$ plus the extra path $\Delta L = \Omega R T^+$ due to the platform rotation. From the equation

$$T^+ = \frac{2\pi R + \Delta L}{c}, \tag{8.5}$$

and the similar one that can be written for T^-, the round-trip time taken by the counter-clockwise beam, the two round-trip times are derived as:

$$T^+ = \frac{2\pi R}{c - R\Omega} \approx T\left(1 + \frac{R\Omega}{c}\right) \; ; \; T^- = \frac{2\pi R}{c + R\Omega} \approx T\left(1 - \frac{R\Omega}{c}\right), \tag{8.6}$$

where it is assumed that $R\Omega \ll c$. By using (8.6), the frequency difference is found to be proportional to the rotation rate of the cavity:

$$\nu^+ - \nu^- = \frac{c}{\lambda}\frac{T^+ - T^-}{T} = \frac{2R\Omega}{\lambda}, \tag{8.7}$$

where λ is the unperturbed laser wavelength.

Thus by measuring the frequency difference, the rotation of the laser cavity (and any vehicle to which it is mounted) can be determined. The ring laser gyroscope is currently utilized in the inertial navigation systems of military aircrafts, commercial airliners, ships and spacecrafts.

8.2.3 Optical Fiber Sensors

Most optical sensors are based on the following principle: if the refractive index of a material depends on an applied external field, then the amplitude of the external field can be determined by measuring the change of phase delay suffered by a light beam that is traveling across the material. Since the refractive index of a material can depend on temperature or pressure or electric field (Pockels effect) or magnetic field (Faraday effect), it becomes possible to design optical sensors measuring each of these fields. Particularly interesting, because of their high sensitivity and design simplicity are optical fiber sensors. In this section, a few examples are considered with the aim of illustrating the possibilities of this approach.

Temperature and strain sensors. In principle, a measurement of phase delay requires an interferometer, such as the Mach-Zehnder interferometer, described in Chap. 4, or the Michelson interferometer, shown in Fig. 8.3. A much simpler scheme can be utilized with optical fibers by exploiting the properties of the fiber Bragg gratings (FBGs) described in Sect. 6.5. FBGs reflect only the wavelength of light that corresponds exactly to the periodicity of the grating. The Bragg wavelength, as expressed by (6.37), can be shifted by changing the grating period or the effective refractive index of the grating. Strain modifies the grating period. Temperature variations affect both the refractive index and the grating period, but around room temperature the effect of temperature on refractive index is about one order of magnitude larger than that of thermal expansion (or contraction). The typical response of the Bragg wavelength shift to strain is about 1 pm/$\mu\varepsilon$ in the near infrared. The unit of strain is ε: 1 $\mu\varepsilon$ corresponds to the elongation by 1 μm of a 1 m long fiber. The typical temperature response is about 10 pm/°C. The Bragg sensor can be interrogated by introducing a broadband light beam and monitoring the back-reflected radiation with a spectrometer that includes an array of CMOS photodetectors, as shown in Fig. 8.6. There are a number of situations where it is necessary to monitor a parameter in many points along a length. Optical fibers can be installed over a measurement path in a large structure, like a dam, a bridge, an aircraft to monitor the spatial and temporal behavior of strain and temperature, with the fiber having several

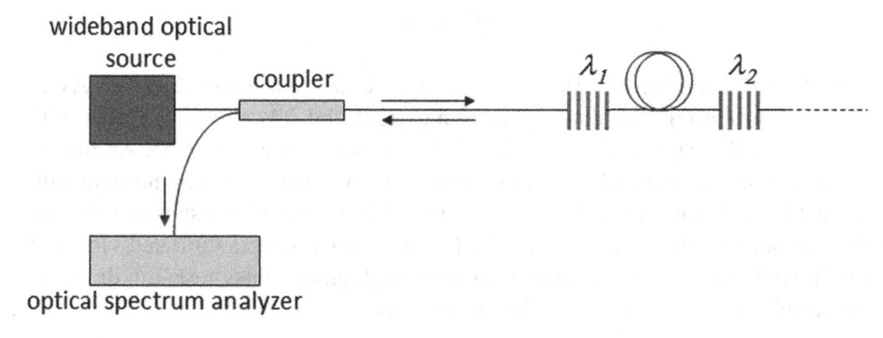

Fig. 8.6 Fiber Bragg grating sensor

FBGs spaced at intervals along its length. If each grating has a different pitch, then it becomes possible to monitor the environment along the whole fiber because each reflection can be recognized by its unique wavelength.

Temperature sensing distributed over a fiber length of several kilometers can also be performed by sending a short laser pulse along the sensing fiber and detecting the backscattered Raman light with high temporal resolution. The power ratio between anti-Stokes and Stokes light, expressed by (7.59), depends on the fiber temperature, but it is independent of the laser power, the launch conditions and the composition of the fiber (provided no fluorescence occurs). The only correction required is a small allowance for differences in the fibre attenuation at the Stokes and antiStokes wavelengths, when long lengths of fiber are used. By using a pulsed semiconductor laser emitting at 1550 nm (the minimum loss wavelength for silica fibers), the Stokes and anti-Stokes lines are found around 1650 and 1450 nm, respectively. The temperature distribution over a few tens of kilometers can be derived from the measured power ratio with a spatial resolution of a few meters and a temperature resolution of a few degrees Kelvin. Stimulated Brillouin scattering may also be used for distributed temperature sensing. The sensing capability here arises from the dependence of the local acoustic velocity in glass on both strain and temperature. As shown by (7.65), the Brillouin frequency shift is inversely proportional to the acoustic velocity. The technique is more complicated than the case of the Raman sensor, because the measurement of the Brillouin frequency shift requires a pulsed pump and a counter-propagating continuous-wave probe. Positional information is obtained by a time-domain analysis of the fiber output.

Electric current sensor. This type of sensor is based on the Faraday rotation effect, described in Sect. 2.5.4, which states that the polarization of light waves is rotated when propagating along (or opposite to) a magnetic field inside a magneto-optic material. The sensor uses the propagation of a linearly polarized optical beam in a bulk material or in an optical fiber exhibiting a non-zero Verdet constant. Compared with the bulk devices, the Verdet constant of optical fibers is quite small, but their optical path length can be increased to compensate for this, by winding the fiber with a large number of turns around the current conducting element. The intensity of the electric current i is obtained by measuring the rotation angle θ and using (2.153):

$$\theta = C_V \mu_o N i, \tag{8.8}$$

where N is the number of turns and C_V is the Verdet constant of the fiber. As an example, considering a cable that carries a current of 1 kA, choosing a fiber with $C_V = 4\ \text{rad}(\text{T}\cdot\text{m})^{-1}$ and putting $N = 20$, it is found that $\theta \approx 5.6°$. An important feature of this approach is that the optical measurement does not interfere with the electric configuration of the circuit, as would be the case for a standard electromagnetic sensor. This fiber sensor, which can be made temperature and vibration insensitive, is especially well fit to measuring the high currents present in the high-voltage substations of power distribution systems.

8.3 Laser Materials Processing

When an intense laser beam is even partially absorbed by a material, the absorbed energy goes into heating. The local temperature may become high enough to liquefy or vaporize the material. The time it takes to raise the temperature can be so short that material is melted or removed before the heat can diffuse out from the point of beam contact. It is then possible to perform materials processing operations, such as cutting, welding, drilling, and marking. Essentially, all materials are processable, metals, plastics, glass, semiconductors, textiles, leather and cardboard. The materials processing market is the second most important photonic market, after the information and communication technology.

Lasers are very versatile sources for materials processing. The main advantages of the laser as a working tool are the following:

- Since there is no physical contact with the processed material, the laser tool does not wear out, and there is also a reduced probability of a contamination.
- There is a variety of available laser wavelengths, and thus it is always possible to find the appropriate wavelength of operation for any type of material.
- The beam intensity can be easily modulated spatially and temporally. This allows the integration of monitoring and control systems based on intelligent photonic sensing techniques.
- Laser fabrication is environmentally attractive because it reduces energy consumption and chemical waste.

The laser sources most utilized for materials processing are the following; the diode-pumped neodymium laser and the ytterbium fiber laser, both emitting in the near infrared around $\lambda = 1\ \mu\text{m}$, the CO_2 laser and the Er-YAG laser, both emitting in the mid infrared (9–10 μm), and the excimer laser, which emits in the ultraviolet. In some applications where intense green, blue, or ultraviolet light is needed, the emission of a near infrared laser is frequency-doubled or tripled, by using the nonlinear processes described in Chap. 7.

Laser materials processing covers a large variety of applications, including the possibility of machining materials that would otherwise be very difficult to process with conventional tools. In the following some of the most important applications are briefly described.

- Solid-state and fiber lasers operating on a single transverse mode produce a Gaussian beam that can be focused to the smallest of spot sizes. The huge power densities associated with the small spot produce excellent cuts in thin and thick metals, and deep penetration for welding applications. Lasers are widely used for the process of scribing, in which a pulsed laser is used to drill a sequence of blind holes along designed fracture lines. Examples are the scribing of silicon solar cells and of ceramic substrates for the electronics industry. Laser cutting processes are extensively used in the industrial production of plastics, fabrics, and paper.
- Multi-mode lasers offer scalability of powers up to tens of kilowatts, while still maintaining a very good beam quality. The wide range of applications with these

lasers include cutting of thick automotive parts, heat treating industrial components, welding titanium panels for aerospace industry, deep penetration welding for ship building industry, and clean cutting of thick stainless steels.
- Marking is the process through which an industrial product is identified with a code, a series number, a date, a producer name, and so on. Q-switched solid-state and fiber lasers offer excellent performance in laser marking applications. The ability to focus the pulse power to small spot sizes enables high-speed, high-quality marking on metals and almost any kind of surface.
- Additive manufacturing, also called three-dimensional printing, is a new process of making a three-dimensional solid object of virtually any shape from a computer-aided-design (CAD) model. In this process, successive layers of material are laid down in different shapes, and a high-power laser (a CO_2 or an ytterbium fiber laser) is used to sinter or melt powder granules of a material to create fully dense solid structures.

8.4 Biomedical Applications

The biomedical applications of photonics can be broadly divided into those based on high-intensity processes, in which the laser beam melts or vaporizes the biological material, and those based on low-intensity processes, in which the laser beam is a diagnostic or manipulating tool. The interaction of light with biological materials produces a variety of different effects. Thermal effects arise from the conversion of the energy of absorbed light into heat through a combination of various non-radiative processes. These do not show a strong wavelength dependence. For high-intensity of illumination, the temperature at the laser spot on the biological tissue can become large enough to cause melting or coagulation. Another effect, which can be induced by ultraviolet light, is photoablation. Since the energy of an ultraviolet photon is larger than the energy of some molecular bonds, cellular components can be decomposed, and thus photofragmented species are released from the tissue. Thermal effects and photoablation are the basis for laser surgery.

The area of biomedical applications of photonics, especially when the emphasis is on low-intensity processes, is now frequently called biophotonics. An effect observed when molecular constituents of cells and tissues are excited to a vibrational-electronic (vibronic) level by the absorption of a photon is fluorescent light emission. Fluorescence is a process by which an excited molecule returns to the ground state in two steps. Firstly, by quickly releasing in a non-radiative way its vibrational energy and secondly by emitting a photon with an energy smaller than that of the absorbed photon. Fluorescence has an important role in biophotonics, as a means to detect the presence of specific molecular groups, or to monitor the interactions of a molecule with its environment. Also, as will be discussed below, the main optical bioimaging methods are based on fluorescence.

In some cases, instead of emitting a photon, the molecule excited to an upper energy level uses its excitation energy to start a chemical reaction. Such a photochemical reaction is the basis, for example, of the procedure called photodynamic therapy, in which light is used to activate a molecule (called photosensitizer) that then triggers a chain of chemical reactions leading to the destruction of a cancerous or diseased cell.

The use of photonics for diagnostics is important for biological research, and may also lead to the development of easy-to-access, minimally invasive, low-cost screening methods. One aim is to provide a means of early detection of diseases. As an example, "in vivo" studies utilize optical-fiber endoscopy for the examination of tissues inside the human body in order to spot the presence of tumors. Another topic of biophotonic interest is trapping and manipulation of biological cells by using laser Gaussian beams.

The following sections briefly treat the applications of photonics to ophtalmology and bioimaging.

8.4.1 Ophtalmology

The first biomedical application of lasers, dating back to the early 1960s, was in ophtalmology, concerning the cure of retinal detachment. This kind of treatment is still widely utilized. Another very popular application is the correction of nearsightedness, or myopia.

Retinal detachment. The retina is a light-sensitive layer of tissue, positioned on the inner surface of the eye. Light absorbed by the retina initiates a cascade of chemical and electrical events that ultimately transmits electrical signals to the brain through the optic nerve. A retinal detachment is a separation of the retina from the underlying tissue within the eye. Retinal detachment is one of the leading causes of blindness. Green laser light can travel through the eye lens and the vitreous humor without being absorbed. The laser beam can be focused onto the retina, selectively treating the desired area while leaving the surrounding tissues untouched. This approach exploits the capability of the laser to perform the welding operation in a position that is not accessible to a mechanical tool. Retinal detachment is treated by using an argon laser ($\lambda = 515$ nm) or a frequency-doubled neodymium laser ($\lambda = 532$ nm). These wavelengths are close to the sensitivity maximum of the eye, and thus are very effectively absorbed by the retina.

Correction of myopia. Several visual defects, such as myopia or astigmatism, are due to the fact that the eye lens does not correctly focus images at the retina position. A cure for myopia involves using a laser to photo-ablate small sections of the cornea, which is a thin layer of tissue covering the front of the eye, to correct its radius of curvature so that light is less tightly focused through the eye lens and onto the retina. The operation is performed by using an excimer laser, emitting ultraviolet light. In this type of treatment, three main procedures are followed, known as photorefractive keratectomy (PRK), laser epithelial keratomileusis (LASEK), and laser in situ keratectomy (LASIK).

Artificial retina. A revolutionary application that is presently at an early stage of development is the realization of an artificial retina. This involves the effort to develop an implantable microelectronic retinal prosthesis that restores useful vision to people blinded by a retinal disease, such as diabetes, that makes inactive the light-sensing cells in the retina. The device consists of a multi-electrode array in a one-inch package that allows the implanted electronics to wirelessly communicate with a camera mounted on a pair of glasses. It is powered by a battery pack worn on a belt. The device is intended to bypass the damaged eye structure: the array provides the electrical signals which are transmitted to the brain through the optic nerve. This is a technology that combines microelectronics, neurobiology and photonics.

8.4.2 Bioimaging

Some recently developed optical methods can be applied to the observation of a wide range of biological objects at high resolution.

Fluorescence. In addition to traditional optical microscopy, bioimaging methods based on fluorescence have become increasingly important. A number of cellular constituents fluoresce when excited by visible light. This fluorescence is called endogenous fluorescence. When no endogenous fluorophores are present, as it happens, for instance, with DNA, labeling of biological structures with exogenous fluorophores is needed for bioimaging. A very important fluorophore is the green fluorescent protein (GFP) which is not naturally present in most cells, but can be generated in situ by a specific biochemical method. The GFP has a major excitation peak at a wavelength of 395 nm and a minor one at 475 nm. Its emission peak is at 509 nm, which is in the lower green portion of the visible spectrum. By using different fluorophores specifically targeting different parts of the cell and by optically filtering the fluorescent emission, it is possible to visualize separately the different parts of a specimen. By using a scanning optical microscope full three-dimensional images can be reconstructed.

High-resolution fluorescence. Further increased spatial resolution can be obtained by two recently developed nonlinear optical methods, known as two-photon confocal microscopy and stimulated emission depletion (STED) microscopy. In two-photon microscopy a fluorophore is excited by simultaneously absorbing two photons, each one having half of the energy required to bring the molecule to its excited state. Two photon absorption, mentioned in Sect. 7.4, is a nonlinear process with a small cross-section. Since the transition probability for simultaneous two-photon absorption is proportional to the square of the instantaneous light intensity, as shown by (7.45), a train of mode-locked ultrashort laser pulses with a large peak power ($\approx$50 kW) and a low average power is utilized. The average power must be kept very low to minimize any thermal damage of the biological specimen. In two-photon microscopy the excitation beam is tightly focused, and thus the region outside the focus has little chance to be excited. This eliminates out-of-focus fluorescence, so that the excitation region is strongly localized. The whole sample can be sectioned by scanning the focal point. Two-photon microscopy can use a near-infrared-wavelength laser as the excitation source and produce fluorescence in the visible range. Using infrared photons instead of ultraviolet or blue photons gives the advantages of less photo-damage, and better penetration into the sample. The spatial resolution can be smaller than the size of the focal region of the exciting beam. STED functions by deactivating the fluorescence of specific regions of the sample while leaving a center focal spot active to emit fluorescence. This is achieved by using a double-laser design, with a doughnut shaped depletion beam. By the STED method a spatial resolution of the order of 10 nm can be attained.

Endoscopy. An endoscope is a medical imaging device inserted into the patient's body to get a clear view of the inside. Endoscopy is a great medical breakthrough because it helps to reduce diagnostic surgeries. In a fiber-optic endoscope light is guided to the area under investigation by non-coherent fibre optic bundles (bundles where the optical fibres are not lined up at both ends). The image is transmitted back by a coherent fibre optic bundle (a bundle where the optical fibres are lined up at both ends of the fibre so that an image can be transmitted).

Optical coherence tomography. Optical coherence tomography (OCT) is a reflection imaging technique similar to ultrasound imaging. A light beam (usually in the near infrared) is back-scattered from the biological tissue under study, in order to produce an image. OCT is based on an interferometric scheme similar to that illustrated in Fig. 8.3. In general terms, the phase relation between successive wavefronts of the optical signal is maintained within a distance known as "coherence length", l_c, which is given by the velocity of light divided by the bandwidth $\Delta\nu_s$ of the source. The interferometer produces a well-defined interference pattern only if the path-length difference between reference beam and back-scattered beam is within the coherence length. In the case of OCT the source bandwidth is very broad, so that $l_c \approx 1\ \mu$m. By scanning the path-length of the reference beam, an interference signal appears only in those positions corresponding to path differences within l_c. The interference pattern contains the information about the refractive index variation of the sample, which thus non-invasively provides the optical image.

8.5 Liquid Crystal Displays

Liquid crystal displays (LCD) are found in a variety of electronic devices, including cell phones, tablets, laptop computers, computer monitors, TV screens and digital watches. As a consumer technology, LCD has been in widespread use since the early 1970s when it first appeared in digital watches.

Images in an electronic display are made up of pixels, which are the smallest distinct element in an image. In color displays the pixel is made of three subpixels, red, green and blue, which can form any color, when added together at different levels of intensities. Each individual pixel (or subpixel) is made-up of a light source and an electronic valve which can turn the light on, off, or to some intermediate level. The light valve consists of a thin liquid crystal film, sandwiched between two crossed polarizers. The orientation of the liquid crystal molecules is controlled within each pixel by applying an electric field. To this end, transparent electrodes are patterned on the glass plates of the liquid crystal cell. The electrodes are made of a thin layer of indium tin oxide, a semiconductor that has a good electric conductivity and, at the same time, is transparent to visible light. In the case of the twisted-nematic cell described in Chap. 4, the pixel is bright in absence of applied field, and becomes opaque when the liquid crystal molecules are forced to align parallel to the applied field.

In the case of simple displays, like those of digital watches or pocket calculators, each pixel is driven separately. In the case of television screens, a matrix connection is built: on a side of the liquid crystal panel all horizontal lines are interconnected, whereas at the opposite side the vertical lines are interconnected, in such a way to select a specific pixel by crossing an horizontal line with a vertical line.

At present there are two common methods of back-lighting in LCD flat panels: cold-cathode fluorescent lamp and LED. Fluorescent lamps used to be the most widespread method of back-lighting for LCD TVs, and consists of a series of tubes laid horizontally down the screen. LED backlighting is now very common and has been in use in TVs since 2004.

It is a general property of displays based on polarized light that the viewing angle is limited. Viewing angle relates to how well you can see a display image when looked at from the side of the display. The lateral visibility is greatly improved by using the so-called in-plane switching (IPS), instead of the twisted nematic approach. In IPS the molecular alignment and the orienting electric field are in the plane of the panel.

8.6 LED Applications

LEDs are currently used in a large variety of applications, such as displays, street lights, automotive lighting, ambient lighting, light sources for robotic vision, remote control units, and so on.

LEDs have an important role in displays. They are used for three-color back-lighting liquid crystals displays (see Sect. 8.5), and also for directly producing images in the giant displays at sports stadium, for instance. In those mega-screens, the images must be very bright, in order to be clearly seen even in presence of full sunlight. A pixel inside a mega-screen is about 1 cm in size and includes tens of LEDs of the three fundamental colors. The brightness and color of each pixel are fixed by controlling the bias voltage of all the LEDs making-up the pixel. High brightness LEDs or solid-state lasers are also used for projection displays.

Displays made of organic light-emitting diodes, also known as OLEDs, are now finding more and more uses. Whereas OLED screens were initially limited to the displays of certain cell phones and car radios, they are now entering into the TV market as well. Compared to conventional liquid crystal displays (LCDs), OLED displays can offer enhanced contrast and lower energy consumption. The current LCD architecture is based on rigid sheets of glass and is therefore neither flexible nor lightweight enough for newly developing applications, such as roll-up screens in movie-theater format or large-area light-emitting wallpaper. Printed OLEDs are components where the functional layer is applied to a substrate by means of a printing process. Printing processes are cheaper and can operate with larger formats than the processes typically used in the semiconductor industry.

LEDs are playing an increasing role in general illumination. Lighting by solid-state light sources, concisely called "solid-state lighting" (SSL) has several attractive properties. First of all, LEDs are more efficient than incandescent and fluorescent lamps, and so visible light can be produced with reduced heat generation or parasitic energy losses. Long-life LEDs can provide 50,000 h or more of operation, which can reduce maintenance costs. In comparison, an incandescent light bulb lasts approximately 1,000 h. The best commercial white LED lighting systems provide twice the luminous efficacy (lumens per watt) of fluorescent lighting. Considering that general illumination uses about 20 % of the available worldwide electrical energy, solid-state lighting could offer a significant contribution to energy saving. Colored LEDs are especially advantageous for colored lighting applications because "lossy" filtering of white light is not needed. Another interesting aspect of SSL technology is that the spectral, spatial and temporal distribution of ambient lighting can be readily adjusted, offering new interesting opportunities. LED systems are intrinsically safe, because they use low voltage and are generally cool to the touch.

A last point to mention concerns ultraviolet LEDs based on AlGaN. These sources have the potential to replace mercury lamps inside the sterilization systems, which are used in various industries, such as the food industry, and in biomedicine.

8.7 Photovoltaic Cells

A photovoltaic cell is an electrical device that converts the energy of light directly into electricity by the photovoltaic effect. The working principle is exactly the same as that described for photodiodes in Sect. 5.5.2. A photon absorbed in the depletion layer of a *p-n* junction generates an electron-hole pair. The migration of the photo-generated carriers produces an open-circuit voltage difference across the junction, or a closed-circuit electric current without being attached to any external voltage source. The energy of solar light can therefore be converted directly into electrical energy. The electric signal obtained is continuous, but can be converted into an alternating signal by an inverter. Solar cells were first tested on satellites that were launched in the late 1950s. The tests were very successful, but the cost of those early devices made their utilization for civilian electrical energy generation unpractical.

The motivations for a massive development of electricity generation based on solar energy conversion are quite obvious: (i) the worldwide energy consumption is continuously growing; (ii) at present about 70 % of electrical energy is produced by burning fossil fuels, and this should be lowered because of environmental problems and also, on a long-term basis, because hydrocarbons are a non-renewable source of energy; (iii) solar energy is abundant and is available everywhere on the Earth's surface, so that, in principle, each country (even each community) could become self-sufficient in the production of photovoltaic electrical energy.

The first issue to be discussed in efficient conversion is the selection of the semiconductor material, bearing in mind that the incident photon is absorbed only if its energy $h\nu$ is larger than the bandgap energy E_g. Silicon, exhibiting $E_g = 1.11$ eV (see Table 5.1), can absorb all visible wavelengths. However, if a visible photon is absorbed by silicon, the fraction of photon energy exceeding 1.11 eV is lost to electric conversion, it goes into vibrational energy and is finally dissipated as thermal energy.

Defining the conversion efficiency η as the ratio between the generated electric power and the incident optical power, and considering that visible (solar) photons cover the energy range 1.8–3.1 eV, the upper limit of η for a silicon cell is 30 %. A higher efficiency can be obtained using *GaAs* cells. By a stack of photovoltaic cells made with different semiconductors of decreasing E_g in such a way to cover the whole solar spectrum, the upper limit to η would become 68 %, but at the price of a large increase in cost.

Currently, solar cells are produced from monocrystalline silicon, polycrystalline silicon, and gallium arsenide (i.e., *III–V* solar cells). Although existing monocrystalline or polycrystalline silicon solar cells are economical, their photoelectric conversion efficiencies only range between 15 and 20 %. By comparison, *III–V* solar cells possess a conversion efficiency ranging between 40 and 50 %, but are more expensive. Typically the semiconductor thickness required to absorb 90 % of the incident power is about 30 μm. Second generation solar cells use thin-film technology, which gives a lower efficiency, but is considerably cheaper. Thin-film solar cells can employ not only crystalline silicon, but also amorphous silicon.

Another technology in use is that of a high concentration photovoltaic module, in which the size of the photosensitive area is reduced by concentrating the incident solar power with a lens. Systems containing large-area concentration lenses and small *III–V* solar cells have been successfully tested.

Similarly to the case of light-emitting diodes, where organic compounds can substitute inorganic semiconductors, photovoltaic cells made of organic photovoltaics based on semiconducting polymers can be prepared. This may represent a potentially inexpensive means of generating electricity directly from sunlight. However, at present, the conversion efficiency is lower than that of semiconductor solar cells. The main attraction of organic photovoltaics is that they can be printed at high speeds onto large areas of thin, flexible plastic substrate using roll-to-roll processing techniques. This creates the prospect of being able to coat every roof and other suitable building surface with photovoltaic materials at extremely low cost. Research is ongoing to find organic compounds with a broad absorption spectrum and a sufficiently long chemical stability.

Appendix A
System of Units and Relevant Physical Constants

The international system of units (SI) is used in this book. The three fundamental units are meter (m), second (s) and kilogram (kg). A prefix can be added to each unit to change its magnitude by multiples of 10. A list of prefixes is given in Table A.1.

In engineering it is common to make use of decibel units, abbreviated as dB. Any ratio R can be converted into decibels through the general definition

$$R_o \text{ (in dB)} = 10 \log_{10} R. \tag{A.1}$$

For example, an amplifier gain of $G = 10^4$ corresponds to $G_o = 40\,\text{dB}$. As $R = 1$ corresponds to 0 dB, ratios smaller than 1 are negative on the decibel scale. For instance, if $R = 0.8$, which corresponds to a loss of 20 %, the ratio in decibel is $R_o = -1\,\text{dB}$.

Table A.1 Prefixes of units

Prefix	Power of 10	Symbol
zepto	10^{-21}	z
atto	10^{-18}	a
femto	10^{-15}	f
pico	10^{-12}	p
nano	10^{-9}	n
micro	10^{-6}	μ
milli	10^{-3}	m
kilo	10^{3}	k
mega	10^{6}	M
giga	10^{9}	G
tera	10^{12}	T
peta	10^{15}	P
exa	10^{18}	E
zetta	10^{21}	Z

V. Degiorgio and I. Cristiani, *Photonics*, Undergraduate Lecture Notes in Physics,
DOI 10.1007/978-3-319-20627-1

Table A.2 Relevant physical constants

Speed of light in vacuum	c	2.9979×10^{8}	m/s
Vacuum permittivity	ε_o	8.8542×10^{-12}	F/m
Vacuum permeability	μ_o	1.2566×10^{-6}	H/m
Electric charge of the electron	e	1.6022×10^{-19}	C
Mass of the electron	m_e	9.1094×10^{-31}	kg
Boltzmann constant	k_B	1.3807×10^{-23}	J/K
Planck constant	h	6.6261×10^{-34}	$J \cdot s$

Table A.2 lists the values of the physical constants that appear more frequently throughout the book.

Appendix B
List of Acronyms

Several acronyms are used throughout the book. Each acronym is defined the first time it appears in a chapter. In any case, to offer a further help to the reader, acronyms are listed here in alphabetical order (Table B.1).

V. Degiorgio and I. Cristiani, *Photonics*, Undergraduate Lecture Notes in Physics,
DOI 10.1007/978-3-319-20627-1

Table B.1 List of acronyms

APD	Avalanche photodiode
CCD	Charge-coupled device
CD	Compact disc
CW	Continuous wave
DFB	Distributed feedback
DVD	Digital video disc
EDFA	Erbium-doped fiber amplifier
FBG	Fiber Bragg grating
KDP	Potassium di-hydrogenated phosphate
KTP	Potassium titanyl phosphate
LBO	Lithium triborate
LED	Light-emitting diode
MQW	Multiple quantum well
NA	Numerical aperture
OLED	Organic light-emitting diode
PMD	Polarization-mode dispersion
QCL	Quantum cascade laser
QPM	Quasi phase matching
SH	Second-harmonic
SHG	Second-harmonic generation
SMR	Single-mode reduced
SOA	Semiconductor optical amplifier
SOI	Silicon on insulator
SPM	Self phase modulation
SRS	Stimulated Raman scattering
VCSEL	Vertical-cavity surface-emitting laser
WDM	Wavelength division multiplexing
YAG	Yttriun aluminum garnet

Reading List

General

B. E. A. Saleh and M. C. Teich, *Fundamentals of Photonics*, Wiley Interscience, 2nd ed. 2007.

R. S. Quimby, *Photonics and Lasers: An Introduction*, Wiley Interscience, 2006.

J-M. Liu, *Photonic Devices*, Cambridge University Press, 2005.

J. Wilson and J. Hawkes, *Optoelectronics*, Prentice Hall, 3rd ed. 1998.

Concerning Chaps. 1 **and** 2

E. Hecht, *Optics*, Addison-Wesley, 4th ed. 2002.

R. Guenther, *Modern Optics*, Wiley, 1990.

M. V. Klein and T. E. Furtak, *Optics*, Wiley, 2nd ed. 1986.

J. W. Goodman, *Introduction to Fourier Optics*, Roberts, 3rd ed. 2005.

Concerning Chaps. 3–5

A. Yariv, *Quantum Electronics*, Wiley, 1989.

O. Svelto, *Principles of Lasers*, Springer Verlag, 5th ed. 2010.

W. Koechner and M. Bass, *Solid-State Lasers: A Graduate Text*, Springer Verlag, 2003.

L. A. Coldren, S. W. Corzine, and M. L. Mashanovitch, *Diode Lasers and Photonic Integrated Circuits*, Wiley, 2012.

S. Donati, *Photodetectors*, Prentice Hall, 2000.

Concerning Chaps. 6–8

J. Hecht, *Understanding Fiber Optics*, Prentice Hall, 5th ed. 2005.

R. W. Boyd, *Nonlinear Optics*, Academic Press, 2nd ed. 2003.

G. I. Stegeman and R. A. Stegeman, *Nonlinear Optics*, Wiley, 2012.

G. P. Agrawal, *Nonlinear Fiber Optics*, Academic Press, 5th ed. 2013.

G. P. Agrawal, *Fiber-Optic Communication Systems*, Wiley Interscience, 2010.

S. Donati, *Electro-optical Instrumentation*, Prentice-Hall, Upper Saddle River, 2004.

W. M. Steen and J. Mazumder, *Laser Material Processing*, Springer, 4th ed. 2010.

M. H. Niemz, *Laser-Tissue Interactions*, Springer, 4th ed. 2011.

V. Degiorgio and I. Cristiani, *Photonics*, Undergraduate Lecture Notes in Physics,
DOI 10.1007/978-3-319-20627-1

Index

V. Degiorgio and I. Cristiani, *Photonics*, Undergraduate Lecture Notes in Physics,
DOI 10.1007/978-3-319-20627-1

Printed in the United States
By Bookmasters